Molecular Cloning

Edited by Sadık Dincer,
Hatice Aysun Mercimek Takcı
and Melis Sumengen Ozdenefe

Published in London, United Kingdom

Molecular Cloning
http://dx.doi.org/10.5772/intechopen.98154
Edited by Sadık Dincer, Hatice Aysun Mercimek Takcı and Melis Sumengen Ozdenefe

Contributors
Sadik Dincer, Hatice Aysun Mercimek Takci, Melis Sumengen Ozdenefe, Cyril Couturier, Colin Béatrice, Ann M. Simpson, Alexandra Mahoney, Najah Nassif, Bronwyn A. O'Brien, Endang Semiarti, Yekti Asih Purwestri, Saifur Rohman, Wahyu Aristyaning Putri, Pradip Vitthal Hirapure, Rita Narayan Lakkakul

First published in London, United Kingdom, 2022 by IntechOpen
IntechOpen is the global imprint of INTECHOPEN LIMITED, registered in England and Wales, registration number: 11086078, 5 Princes Gate Court, London, SW7 2QJ, United Kingdom

British Library Cataloguing-in-Publication Data
A catalogue record for this book is available from the British Library

Additional hard and PDF copies can be obtained from orders@intechopen.com

Molecular Cloning
Edited by Sadık Dincer, Hatice Aysun Mercimek Takcı and Melis Sumengen Ozdenefe
p. cm.
Print ISBN 978-1-80355-450-1
Online ISBN 978-1-80355-451-8
eBook (PDF) ISBN 978-1-80355-452-5

Meet the editors

For the past 35 years, Prof. Sadık Dincer has been involved in teaching, research, and academic work in numerous distinguished universities in Turkey. Currently, he is working at the Biology and Biotechnology Departments, Cukurova University, Adana, Turkey. His manuscripts and book chapters have been published in national and international journals and his works has been cited 1167 times. To date, he has trained twenty-six MSc and eleven Ph.D. students. He received the Technology Development Award from the Scientific and Technological Research Council of Turkey (TÜBİTAK) in 2013 and a national study patent in 2019. His research is focused on bacteriology, microbial ecology, industrial biotechnology, and microbial genetics.

Associate Prof. Dr. Hatice Aysun Mercimek Takci received her MSc and Ph.D. in Biotechnology and Biology from Cukurova University, Adana, Turkey in 2007 and 2011, respectively. Since 2009, she has worked at Kilis 7 Aralik University, Kilis, Turkey. Her teaching interests include microbiology, biotechnology, enzymology, microbial genetics, and bacteriology. She has published forty-six manuscripts in national and international journals and her works have been cited 243 times. Her research interests focus on multiple antibiotic and heavy metal resistance in bacteria, production and characterization of bacterial enzymes, bioremediation by bacteria, and aquatic bacterial contamination. Related to these research areas, she has seventeen projects supported by the Scientific and Technological Research Council of Turkey (TÜBİTAK) and is the coordinator of scientific research projects.

Associate Prof. Dr. Melis Sumengen Ozdenefe received her BSc, MSc, and Ph.D. in Biology from Cukurova University, Adana, Turkey in 2009, 2011, and 2014, respectively. During her MSc, she was an international exchange student and a researcher at Anhalt University, Germany for six months. She has been working in the Department of Biomedical Engineering, Near East University, Northern Cyprus since 2014. Her teaching interests include industrial microbiology, bacteriology, biotechnology, enzymology, and environmental microbiology. Her research areas involve enzymes and biosurfactants, which are produced from various bacteria and fungus, for industrial applications, the production and characterization of bacterial enzymes and bacteriocins, the antimicrobial and antioxidant activity of various plant structures, multiple antibiotic resistance, and heavy metal resistance in Gram-negative bacteria. Dr. Ozdenefe's works have been published in national and international journals, conferences, congresses, and symposiums and cited 140 times.

Contents

Preface

Molecular cloning refers to the creation of recombinant DNA molecules. With the discovery of restriction and ligase enzymes, recombinant DNA technology emerged in the 1970s and spurred great advances in molecular biology, particularly in DNA manipulation. This technology, which enables detailed molecular studies to understand the structure and functions of genes, allows researchers to isolate large DNA molecules from different origins, cutting, pasting, reproducing, assembling, recognizing, changing their structure, and recovering them to living organisms. The cloning of genes, which has become a standard laboratory technique, has significantly increased the understanding of gene function in recent years. This book discusses the fundamentals of molecular cloning with chapters on tools for molecular cloning, molecular cloning for medicine, molecular cloning for food and feed, molecular cloning for the environment, molecular cloning methods, and the future of molecular cloning. It is an intended useful resource for graduate and postgraduate students as well as researchers and industry experts in the domains of biotechnology, ecology, enzymology, food engineering, medicine, microbiology, and molecular biology.

Sadık Dincer
Cukurova University,
Adana, Turkey

Hatice Aysun Mercimek Takci
Kilis 7 Aralik University,
Kilis, Turkey

Melis Sumengen Ozdenefe
Near East University,
Nicosia, Turkish Republic of Northern Cyprus

Section 1

Genetic Transformation - Prokaryotic and Eukaryotic Cells

Chapter 1

High Throughput Methods to Transfer DNA in Cells and Perspectives

Colin Béatrice and Couturier Cyril

Abstract

Genome sequencing led to thousands of genes to study and their molecular cloning to provide ORF collection plasmids. The main approach to study their function involves analysis of the biological consequences of their expression or knockdown, in a cellular context. Given that, the starting point of such experiments is the delivery of the exogenous material, including plasmid DNA in cells. During the last decades, efforts were made to develop efficient methods and protocols to achieve this goal. The present chapter will **first give a rapid overview of the main DNA transfer methods** described so far: physical, chemical, and biological. Secondly, it will focus **on the different methods having reached high-throughput nowadays**. Finally, it will discuss **the perspectives of this field in terms of future enhancements**.

Keywords: cell nanoconstriction, cell-penetrating peptide, DNA, electroporation, high-throughput transfection, lipofection, microfluidic, nano-acoustic dispensing, nucleofection, viral transduction

1. Introduction

The most used approach to decipher proteins' function or their interactome is to study the effects induced by the delivery of exogenous materials in living cells (deoxyribonucleic acid: DNA, ribonucleic acid: RNA, oligonucleotides, proteins, and ribonucleoproteins). Coding sequence overexpression, then gene silencing, and genome editing approaches offer a panel of induced biological modifications within cells that allowed us to increase our knowledge of most cellular processes. However, in a post-genome era, thousands of genes must be studied and exogenous material transfer into cells, including DNA, became a limiting factor. Indeed, available technologies predominantly allowed analysis at a gene-by-gene scale, and new approaches were developed to reach higher throughput. Libraries of material such as small interfering RNA (siRNA) [1, 2] and Open Reading Frame (ORF) expressing plasmids collection were developed [3, 4] to cover all proteome. To take advantage of these, concomitant High-Throughput (HT) technologies are pointed out for their transfer in cells. Plasmid DNA (pDNA) transfer in cells (by transfection or transduction) plays a central role when studying the precise biological role of proteins. For pDNAs, several efficient transfection methods were pushed to higher throughput. All these induced changes performed in cells allow not only our understanding on the biological processes of cells' life but also have therapeutic applications [5, 6]. The huge interest in gene and cellular therapy approaches is indeed a motor in the development of highly efficient gene delivery strategies.

In this chapter, we will first give a brief overview of DNA transfer methods in cells, then a more detailed part will focus on those that reached higher throughputs and we will conclude with future expected enhancements.

2. The different methods to transfer DNA in live cells

To promote a biological effect in the cell, exogenous DNA must face several levels of pitfalls starting from the outside of the cell. First, it must cross the plasma membrane composed of a hydrophobic lipid bilayer which naturally prevents hydrophilic material such as DNA from entering the cells. In addition, DNA and the plasma membrane carry a general negative charge that impedes DNA transfer into cells by electromagnetic repulsion. Furthermore, once entered, the DNA has to face degradation mechanisms that occur in the cells. Finally, if part of the exogenous DNA succeeds in passing all these steps, the expected biological effect would be measurable. To circumvent all these, a range of approaches to transfer DNA has been developed. DNA delivery to cells can be divided into three main categories: physical, chemical, and biological methods. Among all approaches, viral ones are the most efficient but present some limitations such as the transgene size and biosafety issues. Physical and chemical methods were developed to circumvent these limitations and are not limited in the size and number of genes to be transferred. Some of these are still largely used whereas some were more a proof of concept. In this section, we briefly describe the physical, chemical, and biological methods.

2.1 Physical methods

As mentioned above, due to the plasma membrane and DNA respective properties, the transfer of DNA into cells is impaired. All the physical methods aim to directly circumvent the hydrophobic and electrochemical repulsion parameters by disrupting the integrity of the membrane and promoting a transient permeability. Physical methods then do not have a limit in cargo size and do not depend on biological mechanisms as a direct material delivery is performed [7].

An evident method is the direct microinjection of DNA in cells, which was performed in early-stage embryo [8], and then human cell lines [9]. It implies micromanipulation of a single cell under a microscope to bypass the membrane barrier using a thin glass needle to inject DNA directly into its cytosol or compartments [8, 10]. This approach is reproducible but tedious due to the need to inject each cell individually.

Electroporation method has emerged when it was shown that an electric field could promote a loss of membrane permeability by transient pore formation [11] thus allowing DNA delivery in cells [12]. DNA, target cells, and electroporation buffer laying between two electrodes are submitted to an electric pulse [13]. This pulse is divided into a high-voltage stage to create temporary pores, and a low-voltage one to allow electrophoresis of DNA through these pores [14]. Extensive optimizations (pulse voltage and duration, buffer composition) were done to balance transfection efficiency with cell viability as the requested high voltage promotes cell death [15, 16].

"Electroporators" devices are nowadays available with many predefined settings to achieve efficient transfection in almost all cell types, even hard-to-transfect ones [17].

Biolistic or micro-projectiles bombarded to the cells represent another delivery mode. The projectiles made of gold-covered by nucleic acids, penetrate into cells by high-speed bombardment [18]. First developed for plants, this approach is also efficient in mammalian cells/tissues [19]. However, the method suffers the cost of particles. Nanoparticles that can bind nucleic acids, and whose small size allows them to pass cell membranes with high efficiency, represent a cheaper alternative [20].

Femtosecond laser optoporation consists in focusing ultrashort laser pulses on a cell membrane to induce a transient perforation. This membrane perturbation allows the pDNA transfer [21]. Many cell types can be transfected using a variety of laser sources [22, 23]. Despite efficient, due to the needed laser focusing on a single cell level, its throughput is limited.

Acoustoporation or sonoporation uses ultrasounds to induce a transient plasma membrane disruption promoted by bubbles cavitation phenomenon and thus allowing gene transfer [24, 25]. The method was enhanced by the use of high-frequency waves creating reversible nanopores and furthermore promoting "molecular bombardment" on the bilayer membranes that enhances DNA delivery while limiting cell mortality [26].

Passing constriction or nano-constriction is an approach based on the mechanical deformation of cells as they pass through micro constrictions channels [27]. This controlled compression induces transient pores formation into the cell membrane and allows DNA entry from the surrounding buffer [27]. This method is expected to be universal and showed efficiency in easy and hard-to-transfect cell lines like primary and stem cells [28].

The last method, magnetofection, has been classified as a physical method. It is based on magnetic nanoparticles (MNPs) coated with transfection reagents that bind nucleic acids and promote cell entry [29]. Indeed, MNP only induces the concentration of the MNP on the cells mat when a magnetic force is applied but is per se not able to transfer DNA into cells. However, it enhances DNA delivery up to several hundred and allows to lower DNA consummation, and is furthermore efficient in hard-to-transfect cells [30, 31].

2.2 Chemical methods

Interest to develop non-viral and reproducible gene delivery methods has led to the use of chemical reagents. Chemical transfection methods represent an alternative way to bypass the membrane barrier and furthermore try to protect DNA from degradation within cells [32]. These reagents promote DNA compaction, negative charge neutralization, and cell interaction for later entry into cells. These reagents are briefly summarized here after.

Calcium phosphate co-precipitation is the cheapest method and was first described in 1973 [33]. It relies on the formation of a precipitate when the negatively charged DNA binds to calcium ions (Ca^{2+}) [34]. This precipitate interacts with the plasma membrane and enters the cell by endocytosis [35]. This widely used method reaches up to 90% efficiency for easy-to-transfect cells but is impaired by the need for fresh preparation, avoiding any storage of ready-to-transfect plates [36]. The formation of an efficient precipitate depends on several parameters and this method can be toxic for certain cells such as primary ones [37]. Calcium was also shown to enhance gene delivery by other methods [38] and was then tested alone as a transfection reagent (calfection) [39]. The mechanism does not rely on the formation of a precipitate and do not need fresh preparation. Furthermore, the Ca/DNA mixture can be stored for a long period without any loss in efficacy. Intended for batch transfection of the high number of cells, it worked in a 12-wells plate format for adherent or non-adherent cell lines. The easy use, storage ability, and low cost make this method interesting whereas it was not tested so far in higher throughput.

The diethylaminoethyl-dextran (DEAE-dextran) is another reagent that showed efficiency [40]. This polycationic derivate of dextran compacts DNA to form a positively charged complex that later interacts with the plasma membrane to enter cells by endocytosis [41]. The method is simple, low cost, and efficient for many cell types however, new enhanced approaches surpassed it.

Lipofection method is based on the use of lipids and cationic lipids [42, 43]. When mixed with DNA solution, these lipids form liposomes, a kind of vesicular structure with the same composition as cellular membranes and entraps DNA in solution [44]. The formed complexes (lipoplexes) allow DNA delivery through binding to the cell membrane (due to electrostatic forces), cell entry, mainly by endocytosis [45], and release of the DNA for expression. Lipids-based transfection reagents are efficient and mostly insensitive to serum so that medium has not to be removed before transfection. Furthermore, lipofection can be used efficiently in forward or reverse mode transfection in numerous cell lines [46]. Cationic lipids are more and more efficient in DNA delivery, and furthermore efficient on suspension or adherent cells, and for increasing number of cell types, and even hard-to-transfect ones [47].

Cationic polymers are non-lipidic as deprived of a hydrophobic moiety and are then soluble in water. They use a similar mechanism: being positively charged, they interact and compact DNA under the form of polyplexes [48]. They enter the cell by endocytosis, and traffic through endosomes and cytoplasm to finally deliver DNA to the nucleus [49]. This class of reagent has the advantage to limits DNA degradation in lysosomal compartments, increasing delivery efficiency [50].

2.3 Biological methods

Biological approaches to transfer DNA are inspired by natural mechanisms. The most potent of these approaches is gene transfer by viruses. Other methods represent fields in expansion: cell-penetrating peptides or the use of exosomes or vesicular transfer. These approaches do not rely on natural products but on diverted forms to allow the transfer of a gene of interest.

Viral approaches are the highest efficient among all, even in hard-to-transfect cells [51]. To be permissive, the cells must express the receptor interacting with the virus envelope proteins. To enter in almost all cell types, a ubiquitous and widely expressed receptor is preferred. The Vesicular Stomatitis Virus G (VSV-G) protein promotes entry in almost all cell types as the Low-Density Lipoprotein Receptor family is its ubiquitously expressed receptor [52]. Its interaction with the VSV-G protein promotes membrane fusion and allows virus content to be delivered to the cells [53]. The use of a viral vector is however limited in throughput as viral particles have to be produced for each different DNA to transfer. This production involves the cloning of the gene of interest in a viral vector backbone that is later transfected into a packaging cell line to be integrated into pseudo-viral particles. Pseudo-virus are then recovered from the cell's supernatant, concentrated, and titrated before their use for transduction of the target cells. Despite lower throughputs, viral delivery remains the most powerful way to transfer DNA in cells, even in primary cells (90% efficiency).

The fusiogenic envelope G glycoprotein of the VSV-G was also used as a reagent for gene transfer when mixed with plasmid DNA [54]. The resultant product termed "Gesicles" showed 55% transfection efficiency in HeLa cells, and 22% for hard-to-transfect human myoblast cells [55]. Whereas promising, this method did not reach HT yet.

Another interesting biological derivative used for DNA transfer is represented by proteins having natural properties to enter the cells by surface receptors dependent [56] or independent mechanisms [57]. Some natural peptides derived from these proteins, the cell-penetrating peptide (CPP) are able to enter the cell through the membrane [58, 59]. These peptides have short lengths and a global positive charge. Involved mechanisms are still unclear and depend on the CPP (direct penetration, endocytosis, or translocation via intermediate structure in the membrane lipid bilayer). Peptide from the Trans-Activator of Transcription (TAT) protein of the Human Immunodeficiency Virus (HIV) was efficiently used as a DNA carrier in HeLa cells [60]. Some others have been modified and their properties mixed with each other to promote efficient delivery of exogenous nucleic acid into cells [61]. CPP can be engineered by multiplexing peptides with distinct properties or by modifying their composition. One of the engineered CPP is the pepFect14 [62] which showed efficiency for DNA delivery in several cell types such as CHO, HEK293, U2OS, or U87 cells [63].

One last example of naturally occurring biological derivatives is the use of exosomes. First described in 1977, these nano-sized vesicles derived from plasma membrane elements, are involved in mediating messages to proximal and distant cells [64, 65]. This natural process is found in normal or pathological cells [66] and can be turned around to deliver DNA of interest [67].

3. High-throughput batch DNA transfection

To enhance the throughput of the experiments performed on cells transfected by exogenous DNA, it is interesting to do it in a HT way. However, a distinction must be done between experiments performed at HT using transfected cells, and HT transfection of cells. Indeed, depending on assay requirements, transfection of a single condition may be performed using a large volume of suspended cells that are then distributed among several individual wells for subsequent treatments and assays [68]. Alternatively, it is interesting to transfect many different plasmids, each well of transfected cells expressing different transgenes [69, 70]. This difference is generally concomitant with the way the transfection is performed: batch protocol or not.

Batch protocol allows to transfect a large number of cells that are then dispatched in separate wells for further experiments. In this case, all transfected cells in the batch share the same conditions of transfection. This protocol is generally used to limit variability in HT assays for monitoring the effect on a biological parameter under a unique transfection condition. Typically, it can be performed on adherent cells in a forward-protocol modus: cells are plated and transfected 24 h later, according to the transfection reagent's manufacturer instructions. The day or several hours after transfection, adherent cells are suspended and dispatched in multi-well plates (96, 384, or even 1536) for further HT treatments and analysis [71]. Depending on the cells used, the batch transfection is also compatible with the suspended cells that are then directly dispatched on separate wells after transfection.

The batch protocol is not per se a HT transfer of different biological materials in cells, but rather a way to perform HT assays and treatments in separate wells. On the opposite, HT protocols can achieve a true HT transfection in which each well receive a different DNA or transfection conditions.

4. High-throughput transfection protocols

To be able to determine the behavior of cells or biological effects induced by the transfer of many different pDNAs in the cells, a real HT transfection becomes interesting. Several methods allow the management of numerous pDNA or different conditions when transfecting the cells, but to reach HT, good efficiencies are almost necessary. HT transfection can be achieved using several methods that are presented below.

4.1 High-throughput transfection using physical approaches

4.1.1 High-throughput electroporation-based transfection

As described before, electroporation is performed with buffer diluted cells and DNA, subjected to an electrical pulse that promotes membrane destabilization. Many devices, protocols, and dedicated buffers have been implemented to reach universal use. However, it seemed incompatible for HT as each separate transfection must be performed one by one in micro cuvettes. This problem has been solved by the development of new devices able to deliver an electrical pulse simultaneously in several wells on dedicated plates. Harvard Apparatus/BTX developed an up to 96 wells approach using plates with embedded aluminum electrodes. Used with the plate handler Model HT-200, it allows transfection in 8 wells simultaneously and was shown efficient in neurons [72].

Another approach based on an array of 96 suspended electrode pairs fitting on top of standard 96-well plates represents a less expensive approach [73]. Each pair of electrodes can be loaded and held 10–20 μL of transfection mixture by the surface tension. After pulse delivery, the direct addition of the cell culture medium into the array allows the electroporated cells to drop and seed into the underlying microplate. The array is reusable, and uses standard microplates and inexpensive standard buffers, reducing the cost of this approach. In addition, common liquid handling robots can achieve a 96-well transfection time of approximately 1 min. This technology could be adapted to the 384-well plate format using a more sophisticated electrode array design and concomitant robotics.

Whereas successful in almost all cell types, the electroporation method has some limitations. First, it is the most DNA-consuming one of all the HT transfection approaches described so far. Secondly, as a cell suspension is required during the electrical pulse delivery, it avoids its use on adherent differentiated cells mat. Nevertheless, its advantage in terms of success in almost all cell types, and its versatility concerning the material to transfer (not only efficient for DNA) promises electroporation to further future enhancements and use.

4.1.2 High-throughput nucleofection-based transfection

In 2001, Amaxa™, (now owned by Lonza™) launched an electroporation-derived method termed nucleofection as DNA is transferred directly into the nuclei of the cells. It shortens the time of experimentations, by suppressing the necessary nuclear import step of DNA and ensure proper expression of transgene [74]. It relies on an electroporation-based device (Nucleofector) and the use of dedicated buffer solutions to ensure nuclear transfer. The exact mechanism allowing nucleus targeting and buffer composition is kept proprietary. However, since the first published results on natural killer cells transfection [75], it has been widely used in many hard-to-transfect cells with efficiency ranging from 25 to 70% [76]. First, nucleofector devices used nucleocuvettes and were then limited in throughput. New apparatus and dedicated consumables were developed to reach higher throughput: the 96-well Shuttle® device (amaxa AG), in which cells are plated on 96 wells "nucleocuvette plates" and pulsed using Nucleofector™ programs. These plates, made of conductive polymers, allow the current delivery in each well individually. It takes less than 10 min in an automated way to process the entire plate [77]. Many optimized conditions have already been defined using nucleocuvettes depending on cell types (programs for the electrical pulses, cells number, and optimal buffer conditions) and as an advantage, these settings are transposable to nucleocuvette plates [17]. Numerous successful examples have been published ranging from 35 to 70% efficiency: primary chondrocytes [78]; dendritic cells [79], and even H9 hESC [80].

To push further the throughput a 384-well Nucleofector™ requiring 384-well Nucleocuvette™ plates was launched. The complete electrical pulse delivery process takes just one minute, and several wells are processed at the same time [81]. However, the overall process is the same as for the previous model, mixing of cells with buffer and DNA in the wells, delivery of the electric pulse, the addition of fresh medium in the 384-well Nucleocuvette™ for cell recovery, and then their dispense in a cell culture plate for later experimentations.

Whereas versatile and being efficient as electroporation in many cell types, nucleofection is still restricted to suspended cells, impairing its use on morphologically differentiated and adherent cells. Furthermore, the need to transfer transfected cells to a standard plate for further experimentation is a limiting step of the method. The cost of such approaches broadens their wide use in the scientific community due to the price of transfections kits, containing ready-to-use buffers, and nucleocuvette plates.

4.1.3 High-throughput adherent cells electroporation-based transfection

As explained above, electroporation or nucleofection are restricted to cell suspensions. To circumvent this limitation, new kind of electrodes able to deliver the electric pulse on a cell mat was developed.

One of the simplest developed approaches is an electrode device that takes place on top of standard culture cell dishes. The PetriPulser™ (from BTX) consists of 13 gold plated electrodes embedded in an isolating holder placed above the Petri dish containing the cell mat to electroporate [82]. This model fits 35 mm Petri dishes but a scaled-up model, the "Petri dish electrode" made of stainless steel electrodes, fit 100 mm diameter dishes [83]. The 2 mm distance between electrodes is the same as in most cuvettes. A model for transwell cultured cells electroporation: the BTX™ Adherent Cell Electrodes [84] presents a 5 mm distance inter electrodes and may engender adverse effects on cell viability. All these devices are reusable, lowering the cost of this approach that has however not been used so far in published works.

A sophisticated version was launched by Cellectricon™: the Cellaxess®HT. It uses dedicated 384-wells microplates and a capillary embedded microelectrodes array.

Using a platform device, adherent cells seeded in 384-wells plates are washed, electroporated using transfection mixture (loaded from side donor plates), and allowed to recover with fresh medium addition. 96 wells are simultaneously electroporated by the device and throughput of 50,000 wells per day is announced by the manufacturer [85]. However, it was not really used in the academic laboratories as no work has been published except the proof of concept of the manufacturer. They simplified the method by launching the Cellaxes Elektra-Adherent Cell Electroporation System. It is also an electrode-based electroporation system optimized for the *in-situ* transfection of all adherent cell types, which offers superior efficiency and cell viability due to minimal cell processing and the low voltages enabled by the use of capillary electrodes laying above the cell mat. It uses 384-wells plates and delivers the electrical pulses in 96 wells simultaneously thus allowing the rapid management of the entire plate. However, whereas fully automated, the protocol is not homogeneous: some medium must be discarded from the wells to add DNA diluted in the electroporation buffer (Cellaxess Elektra Accelerator Solution). Once the pulse delivered, some fresh medium is added to the wells before returning the plate to the incubator. Such inhomogeneous protocol would render reproducibility harder to achieve. Cellaxess Elektra transfection system allows rapid optimization of the protocol as different pulse protocols can be applied in a single 384-well plate. This approach has not been widely used yet, probably due to the cost of consumables and devices, but was able to transfect primary neurons with an efficiency of up to 50% [86].

4.1.4 High-throughput electroporation-based microarray in situ transfection

Array approaches are based on spotting an array of transfection reagents and material to transfer on a planar slide where cells are later plated. Using such approaches with electroporation method was unimaginable. However, several teams pushed down this restriction by developing custom-made devices to electroporate adherent cells in a microarray manner. Two technologies are suitable for adherents cells: the delivery of the electrical pulse between the bottom and top of chip microwells; or between interlaced microelectrodes laying on the bottom of the dishes under the seeded cells [87, 88].

In the HT *in situ* cell electroporation (HiCEP) method, a microarray electroporation chip composed of 13 × 13 microelectrodes have been developed [89]. The electrodes lay under the cell cultured in a superhydrophobic microwell array chip (SMAR-chip) developed for this purpose. The electrical pulse is delivered simultaneously in the 169 wells, using for each, ten interdigital electrodes covering a 500-μm-diameter area [90]. The approach requires a dedicated platform to assemble the chip before covering it with the cells solution in a Petri dish. Before delivering the electrical pulse, the medium is removed, allowing 24 nL medium nanodrops to stay in each well of the hydrophobic matrix chip. Electroporation buffer is added and rests as nanodrop in each well after aspiration. The material to transfer is deposited by a standard microarrayer, on the top cover slide. Once reversed and placed on top of the wells using a micromanipulator under the microscope, the drops mix with the underlying buffer before electrical pulse delivery. Whereas this method is successful and promising in terms of throughput, it is not affordable for non-specialists, as many skills and specialized materials are required.

Another method was able to electroporate adherent cells, based on a glass gold electrode coated with PEI for pDNA loading [91]. Cells are plated on this electrode and the electrical pulse can be delivered using an additional top cover electrode up to 3 days post-seeding. Transfection efficiency reached 90% in HEK but was also efficient in primary fibroblasts. Although electroporation was performed in 13 mm square areas, this method allowed HT transfection using up to 169 plasmids microarrayed on the electrode. This method seems affordable, as it only requires a gold vaporized electrode.

4.1.5 High-throughput electroporation-based microfluidics transfection

Whereas it remains a field of specialists, microfluidic applications increased in the last decade due to their low-cost advantage, as it can be in-house designed using affordable technologies, and it deals with low quantities of reagents. Microfluidic can manipulate different solutions and mix them, and lead to cell culture and transfection chips design [92]. However, in-house designs might be difficult to reproduce, even more, if highly specialized skills are required. Furthermore, most biological experiments require a subsequent amount of transfected cells, harder to achieve using microfluidic. Despite these limitations, success in microfluidic transfection applications has been published for a wide variety of cells, and even at the single-cell level [93, 94]. First devices lacked the necessary throughput to test numerous transfections conditions in parallel, but recent advances pushed it further. In the field of transfection, two main approaches have been used with microfluidics: electroporation and nano-constriction.

Electroporation in standard 2 mm cuvettes requires high voltage that promotes cell death by a joule heating effect, a local pH change due to water electrolysis, that in turn induces the formation of bubbles promoting cells aggregation and impairing the DNA delivery efficiency [95]. Due to its efficiency, electroporation was used in microfluidic derivatives trying to circumvent some of its limitations. Embedding electrodes in a microfluidic channel can limit adverse effects on cell viability [92]. The diameter of the channel allows the electrodes to be closer to each other's and the use of voltages as low as 1 volt [96], reduces the heating joule effect, electrolysis, and bubbles. pH modifications are still present but enhanced buffer composition improved it [97]. These microfluidics devices mostly use flowing cells transfected in a semi-continuous way [98], avoiding testing many different conditions in parallel and lowering throughput. Some devices allow transfection of adherent cells in micro-chambers using a porous substrate on which cells are seeded. The electric field is then applied through the cells (under/upper compartment). This has been successfully performed on stem cells differentiated in neurons [99]. Despite the latest improvements, microfluidic-based approaches still lack HT. However, due to the booming application of microfluidic, reaching higher throughput would be achievable and a promising way to perform transfection.

4.2 High-throughput transfection using chemical reagents

Most of the chemical transfection reagent allows two kinds of protocols: the forward and the reverse protocol. In forward protocols, DNA and transfection reagent are mixed to form transfection complexes and then distributed on previously seeded cells. Such an approach is harder to manage in a HT way as each different mixture condition implies a different container (tube or wells of multiplate wells) and necessary tedious pipetting steps. However, this kind of protocol can be manually achievable with standard molecular biology material such as multichannel micro-pipettors. An experimented user can transfect one to four 96 well plate manually in 2 h with up to 3 different pDNAs per condition [100, 101]. However, to our knowledge, the forward approach has not been automated so far to reach HT.

The forward mode has been surpassed by the reverse protocol mode. The DNA (eventually with the transfection reagent) is directly dispatched on the final wells (i.e., of a multi-well plate) or a glass slide, and cells are added directly on these deposits. This mode of transfection has several advantages: first it shortens the overall experimental time, second, it can easily be automated allowing to reach HT and good reproducibility. Suitable for such an application, liquid handling devices enable the dispense of low liquid volumes for the multiplexing of different solutions whose concentration and ratio are tightly controlled in each well. Such protocols have been developed for most of the biological material to transfer which includes DNA and follow the technological developments available to do it. An overview of these methods used for DNA transfection in a HT manner is detailed below.

4.2.1 Chemical-based high-throughput transfection

As mentioned before, lipidic transfection reagents are eligible to reverse protocol, making them suitable for potential HT approaches. This reverse mode was shown efficient on CHO cells grown in suspension in a 96 wells-plate format using PerFect Lipids (pFx-6 form Invitrogen) as reagent [100] and even adherent cells using Lipofectamine (Invitrogen). Higher Throughput was reached using Turbofectin8 as reagent (Origene) and plasmids coding 704 different transcription factors dispensed in 384-wells plates [102].

The SMAR-chip described before in the HiCEP method [89], was also applied to HT reverse transfection but using Lipofectamine 2000 as a transfection way instead of electroporation. It allowed the efficient transfection of HEK293 (up to 65% transfected cells) in the 169 wells of the matrix. The authors aimed at producing viral particles using co-transfection of the necessary plasmids with 169 genes of interest. Proper viral packaging and sufficient viral production were shown by successful transduction of side cultured 3T3L1 cells using the supernant of the HEK producing cells.

Tavernier's group reached a much higher throughput in 2002 using reverse transfection for HT transfection of HEK293 cells in its MAmmalian Protein–Protein Interaction Trap (MAPPIT) Arrays approaches to study protein–protein interactions [103]. Effecten reagent was used in a reverse mode protocol to transfect prey expressing plasmids in up to 384-wells plate format using classical liquid handling facilities and a mammalian ORF collection plasmids.

With the emergence of such collection, examples of microplate-based arrays of the huge collection of plasmids have grown. One of the highest throughput was reached using 6049 different human cDNA expression plasmids to study their effect on the promoter activation of the zinc-finger protein RP58, using a luciferase reporter gene [104]. 50 ng plasmid/wells were loaded on sets of 384-well plates and a HT reverse transfection of HEK293 was successfully performed using Lipofectamine 2000.

A HT transfection protocol was reached in 384-wells plates format using non-liposomal polymers (Mirus TransitX2) as transfection reagent [101]. A reverse protocol led to about 90% transfection efficiency (even in cotransfection assay). The originality of this work is the use of a tips-free acoustic delivery of reagent and DNA (Echo nanodispencer from Labcyte™). This device sends multiple droplets of 2.5 nL from a 384-wells source plate to a destination one up to 1536-wells plates. Starting from unique diluted plasmids solutions, the overall process takes less than 20 min for one plate, and transfection ready plates can be stored dry or frozen without loss of efficiency. Cells are seeding on dry or freshly dispensed plates in a reverse mode transfection. The optimized protocol would allow 20,000 human genes transfection in about 18 h on a dedicated automated platform. Nano-quantities of DNA and reagents should render this approach low cost if the nanoaccoustic dispenser was not such expensive. Nevertheless, this protocol renders transfection affordable for newbies as the tedious work of DNAs and reagents combining in each well is controlled by spreadsheet driven software [105].

4.2.2 Chemical-based microarray transfection

In 2001, DNA transfection throughput was pushed further by the use of a microarayer for the generation of transfection ready arrays of DNA [106]. In this study, 140 different plasmids DNA/gelatin mixture were deposited on glass microscope slides as 1 nL spots (of about 150 µm diameter). Effecten, a lipid transfection reagent was used to transfect cells seeded on the overall slide. Each spot led to the transfection of 30–80 HEK cells, in a DNA dose-dependent manner from 10 to 50 pg. Storage of the dried glass slides for more than 3 months did not affect transfection efficiency, allowing the matrix to be prepared in advance of use. Since this princeps study, other groups have successfully used this approach. Using the same reagent, one study transfected 16 different plasmids expressing proteins to study their cellular localization [107]. Another group used this approach for the HT screening of potential therapeutic membrane-displayed single-chain antibodies [108]. A true HT attempt was reached by the use of 1959 un-tagged ORF taken from the Mammalian Gene Collection (MGC) and expressed in HEK cells to identify genes implicated in apoptosis [109]. One similar array approach showed efficiency using Lipofectamine 2000 directly in the DNA mixture before arraying [110]. However, whereas simplified by combining the transfection reagent with DNA before dispensing, it requires about 10-fold more DNA to reach the same efficiency as the above protocols. A throughput of 2880 conditions on a complete 96-wells plate to study v-Src Mutant Protein Function was reached in HEK cells, using 30 spots of pDNAs mixtures per well of 96-wells plate [111]. Lipofectamine 2000 also showed efficiency in another microarray approach testing 600 cDNA spots on a single glass slide using reverse transfection [112]. Authors showed high efficiency in many cell types such as mouse preadipocytes (3T3L1), muscle myoblasts (C2C12), liver hepatoma (Hepa1c1c7), or macrophage (RAW-164.), human cervix epithelia adenocarcinoma (HeLa), or at bone osteosarcoma (UMR-108).

Tavernier's group also pushed further its MAPPIT and MAmmalian Small molecule-Protein Interaction Trap (MASPIT) microplate-based array to microarrays using attractene (Qiagen), a non-liposomal lipid, as transfection reagent and a fluorescent reporter gene in place of the initial luciferase reporter [113]. Here, the ORFeome derived prey plasmid collection (15,000 cDNA) and a fluorescent reporter plasmid was mixed in 384-wells plates used as a matrix for further depositing by a microarrayer on polystyrene plates, to reach an industrial scale.

All these arrays' methods are impressive in terms of throughput as many conditions, or different expressed genes, can be tested simultaneously in parallel cells. However, they require a consequent preparation time. DNA dilution, most of the time with gelatin, and optimally with the transfection reagent are generally performed in 96 or 384-wells plates. Once done, an arrayer robot is then plunging its tips for deposition of the DNA on several slides. The tips must be washed with detergent and then sonicated or heated to avoid cross contaminations before arraying the next DNA mixture. Finally, when the full array is printed, the slides have to be dried for 12 h to 2 days before later use and cell seeding. At the end of the experiment, a slide scanner became necessary to analyze transfected cells. The real throughput of such methods is then truly high once the arrays are ready to be incubated with the cells. Once the reagent used is efficient with the cell type requested, the throughput becomes only dependent on the liquid handling facility available in the lab. However, the method needs a certain financial investment for robotics, microarrayer platform and a scanner as the spots size and inter-distance need high resolution scanning to be analyzed.

4.2.3 Chemical-based high-throughput microfluidic transfection

As previously mentioned, microfluidic is now widely used due to the miniaturized scale it allows. Whereas it was applied to transfection using nanoconstriction or electroporation, it can also be used as a liquid and cell manipulation tool to perform transfection using chemical reagents. Schudel et al. first developed an inexpensive microfluidic-based miniaturized RNAi screening platform [114]. It relies on the use of a lipid-based transfection mixture and is low throughput as a maximum of 8 parallel transfections can be performed on this chip.

In another study, a two microchannel irrigating 8-chambers was designed on a glass slide [115]: 10 nL of a reverse transfection mixture containing gelatin, fibronectin, Lipofectamine 2000, and plasmid DNA were arrayed on a coated glass slide. This slide is mounted under the microscope, to face the microfluidic embedded chambers and showed successful transfection of Cancer LBT-N2b cells with almost no induced mortality, but the throughput was still clearly limited.

Based on the same kind of chambers design, the highest throughput was reached with a microfluidic chip of 1.6 × 5.8 cm containing 280 separate chambers. In about 10 min, the complete chip is loaded with about 600 cells per chamber of 500 μm diameter [116]. A set of valves allows the loading of different cell densities or even cell types. Once cells are loaded, the functional chip is obtained by alignment of the chambers to 280 DNA arrays (Lipid-DNA transfection mixtures) spotted on polylysine matrixes in an automated manner (about 2 h to complete). The assays showed a high transfection rate (99% efficiency) using an optimized condition but a cell line-dependent optimization is necessary. Whereas feasible, microfluidic managed reverse transfection still seems to have a long road to meet the scientific community mostly due to its required skills in the field, to be able to reproduce or use such devices.

4.3 High-throughput transfection using biological derivatives

As discussed before, some natural biologicals materials, viruses, proteins, peptides, or macromolecules have shown cell-penetrating properties and their ability to deliver different molecules to target cells either in their natural form, modified, and sometimes multiplexed by engineering. Here are some examples of such approaches that reached HT in the delivery of DNA into cells.

4.3.1 High-throughput DNA transfer using viral approaches

The main limiting factors to reach HT with viral delivery is the ability to produce these particles (i.e., biosafety cabinets class 2 or 3), in a HT manner (one independent viral production for each cDNA to transduce), and at a sufficient titer to promote efficient transduction of target cells. This production step has been shown feasible at HT in a pilot study with 1990 ORFs from the mammalian ORFeome collection [117]. In an automated platform, HEK cells were reverse co-transfected with these "gene of interest" plasmids and viral packaging ones to allow the production of the corresponding lentivirus in a 96-wells plate format (viruses transferring one ORF per wells). Supernatants (cDNA containing viral particles) were used to transduce target cells seeded in 96-wells plate format. In a similar manner, up to 16,000 cDNA were pushed to HT lentiviral production in 96-wells plates for later HT expression in target cells [118].

The previously developed SMAR-chip [89] was also used for viral particles production on the 169 matrixes embedded microwells, using reverse lipofection [119]. The method showed sufficient production to transduce 3T3L1 cell cultured in parallel to the producing cells array.

Despite these advances, such methodologies remain difficult to settle routinely due to the required material, specific skills, knowledge, and adequate biosafety facilities. To render it more accessible, some companies now propose ready-to-use kits in 96-wells microplate format to produce viruses in high titers from lab collection of cDNA [120]. Despite these limitations, this approach is highly promising as being universal for almost all cell types with high efficiency and furthermore efficient on suspension or adherent differentiated cells.

4.3.2 Protein-derivatives based HT transfection

One example of protein derivatives used is collagen derivatives, which are produced by collagen treatment or digestion. Atelocollagen, is a polymer obtained by pepsin treatment of type I collagen that shows various effects in cell and animals. Atelocollagen condenses and delivers DNA, antisense oligodeoxynucleotides, or siRNAs into cells on its own [121]. Protocol based on this polymer reached a HT microplate array level in 2001, with a collection of pDNA showing a long-term gene expression in HEK cells [122]. The array can handle long storage without loss of efficiency. Another study reached HT transfection in PC-12 cells using Atelocollagen and 288 different plasmids dispensed in 96-wells microplate arrays [123]. The advantages of these last approaches remain in the fact that atelocollagen intrinsically regroups two properties in a single bio-product: DNA condensation and cell entry of the formed complexes into cells. Furthermore, it is derived from a biocompatible natural material and per se is rarely cytotoxic for cells.

4.3.3 High-throughput cell-penetrating peptides-based transfection

Due to their potential, the use of CPP was pushed to HT transfection. The surface transfection and expression protocol (STEP method) is the only biological derivatives-based DNA transfer approach that reached such HT. It relies on the use of transferrin receptor, polylysine, adenoviral penton protein, and the HIV Tat protein to engineer some chimeric proteins. These combine functional motifs: binding of the DNA, binding to cell-surface receptors, the facilitated passage across membranes, the DNA targeting to the nucleus, and also adhesion and survival of the target cells on the arrayed spots [124]. The DNA/recombinants proteins mixtures are loaded in 384-wells source plates for standard arraying. Optimized conditions showed efficient GFP plasmid transfection efficiency (50–80%) and transgene expression in several cell types from easy to transfect HEK cells to more difficult ones such as SH-SY5Y neurons, N2A neurobalstoma cells, or PC-12 pheocromocytose cells. This method is promised for future enhancements accompanying the study of new CPPs. Indeed, many CPP have already been identified and validated leading to the creation of a dedicated database in 2012 referencing 843 CPP identified so far [125]. However, an exhaustive list is impossible to give as some are still identified nowadays and the developed database now contains 1700 unique CPPs 10 years later [126]. Some of them may represent better candidates for DNA transfer. This DNA transfection approach is also of great interest for gene therapy as it enables a kind of transduction, efficient like viral particles but without all the safety concerns for their production and use [127].

5. Conclusions and perspectives

Last human genome sequencing assembly led to more than 24,000 genes to study [128]. Many approaches to transfer DNA in cells were then pushed to HT to interrogate each gene function. While still in progress with developments of new reagents and methods, HT DNA transfer approaches are already available. The main remaining challenge is to render them cheaper and affordable for non-specialists.

Among physical approaches, electroporation methods surpass the others being efficient in all cell types. The suspended electrodes array design represents several advantages: it is low cost, usable with standard electroporators and liquid handling devices, but is currently limited to 96-wells plate format [73]. Due to the technical design, it should be amenable to a 384-wells plate format. However, it is still restricted to suspended cells. Electroporation approaches for adherents' cells have also been developed in 384-wells plate format, but suffer from their cost and their need for expensive consumables [86].

Microfluidics devices suffer from the required skills and technologies to be assembled and used. Microfluidics combined with electroporation appears as a solution to some limitations but chamber-based devices seem too far from the standard assays format to be widely used. Applicable to all cells, microfluidic devices based on semi-continuous electroporation of flowing cells currently lack the necessary throughput [98]. The same concern is pointed out for nanoconstriction-based transfection designs [28]. However, higher throughputs would be amenable as microfluidic manipulation of cells and solutions in an automated way is possible at a high rate. Such a device would advantageously require an automated loading of pDNA from a source plate to the chip, transfection of the expected amount of cells and their dispensing on microplate wells, and then a rinsing step of the chip before starting a new cycle with the next pDNA. Indeed, these technologies are readily available and just need to be combined [129].

Methods combining microfluidic electroporation and DNA arraying seem at that time more difficult to be widely used. Indeed, many skills are necessary to prepare the functional chip: design of microfluidic device, micro arraying of the DNA, and even a micromanipulation platform to mount the complete functional chip [116]. This and the cost of the required material will limit its use in the scientific community.

Chemical-based transfection is readily available and represents the methods that reached the highest throughputs. The reverse protocol is the preferred mode with the use of lipids or cationic polymers and achieved a throughput of several thousand independent points [104]. A major limitation is that transfection occurs after a suspension step when cells are seeded. The use of the same approaches but in a microarray manner, also showed HT being possible to perform transfection on adherent differentiated cells. However, in this case, the use of microfluidics and their inconvenients impair its wide use.

Biological approaches also reached HT. Viral transduction is the most powerful tool to transfer DNA. However, biosafety concerns, and furthermore difficulties to produce viruses in arrays format avoid its wide use. CPP-based delivery is of great potential and a more important use should be expected in the next decade with the advance of our knowledge in this field.

In order to deliver an easy way to perform transfection even by novices, a fully automated transfection protocol was developed using a tipless nano-acoustic dispenser device [101]. Users just have to indicate amounts of DNA and transfection reagent to be delivered in each well using a custom spreadsheet and prepare the requested source plate. The device-controlled software performs the tedious dispensing from the source plate to destination one, based on the spreadsheet [105]. The method could be applicable to any chemical reagents and even to CPP-based approaches. This approach could also be performed in forward mode then allowing adherent differentiated cells transfection. Newer versions of the device allow 1536-wells plates as the source and can now dispense in 3456-wells plates. It then becomes possible to regroup the human ORFeome collection plasmids on less than 15 sources plates, and their transfer to about seven 3456-wells plates only. The method allows preloading of the plates and long-term storage before cell dispensing. However, the cost of the dispenser is extremely huge and still impairs its use. The future end of the patented technologies protection, expected in 2025–2030, should induce a price drop due to competitors' and wider the use of such an approach.

Acknowledgements

The authors thank the National Institute of Health and Medical Research (INSERM) for its financial support for the publication of this chapter.

Author details

Colin Béatrice[1] and Couturier Cyril[2*]

1 LIttoral, ENvironment et Sociétiés (LIENSs), UMR CNRS 7266, La Rochelle University – BCBS Group (Biotechnologies et Chimie des Bioressources pour la Santé), La Rochelle, France

2 INSERM, Institut Pasteur de Lille, U1177—Drugs and Molecules for Living Systems, University of Lille, Lille, France

*Address all correspondence to: cyril.couturier@univ-lille.fr

References

[1] Kaykas A, Moon RT. A plasmid-based system for expressing small interfering RNA libraries in mammalian cells. BMC Molecular and Cell Biology. 2004;**5**:1-11

[2] Sen G, Wehrman TS, Myers JW, Blau HM. Restriction enzyme–generated siRNA (REGS) vectors and libraries. Nature Genetics. 2004;**36**:183-189

[3] Bechtel S, Rosenfelder H, Duda A, Schmidt CP, Ernst U, Wellenreuther R, et al. The full-ORF clone resource of the German cDNA consortium. BMC Genomics. 2007;**8**:1-12

[4] Lamesch P, Li N, Milstein S, Fan C, Hao T, Szabo G, et al. hORFeome v3.1: A resource of human open reading frames representing over 10,000 human genes. Genomics. 2007;**89**:307-315

[5] Pushparaj PN, Aarthi JJ, Manikandan J, Kumar SD. siRNA, miRNA, and shRNA: In vivo applications. Journal of Dental Research. 2008;**87**:992-1003

[6] Modell AE, Lim D, Nguyen TM, Sreekanth V, Choudhary A. CRISPR-based therapeutics: Current challenges and future applications. Trends in Pharmacological Sciences. 2022;**43**:151-161

[7] Wells DJ. Gene therapy Progress and prospects: Electroporation and other physical methods. Gene Therapy. 2004;**11**:1363-1369

[8] Capecchi MR. High efficiency transformation by direct microinjection of DNA into cultured mammalian cells. Cell. 1980;**22**:479-488

[9] Diacumakos EG, Holland S, Pecora P. A microsurgical methodology for human cells in vitro: Evolution and applications. Proceedings of the National Academy of Sciences of the United States of America. 1970;**65**:911-918

[10] Lin TP. Microinjection of mouse eggs. Science. 1966;**151**:333-337

[11] Benz R, Beckers F, Zimmermann U. Reversible electrical breakdown of lipid bilayer membranes: A charge-pulse relaxation study. The Journal of Membrane Biology. 1979;**48**:181-204

[12] Wong T-K, Neumann E. Electric field mediated gene transfer. Biochemical and Biophysical Research Communications. 1982;**107**:584-587

[13] Fajrial AK, He QQ, Wirusanti NI, Slansky JE, Ding X. A review of emerging physical transfection methods for CRISPR/Cas9-mediated gene editing. Theranostics. 2020;**10**:5532-5549

[14] Shigekawa K, Dower WJ. Electroporation of eukaryotes and prokaryotes: A general approach to the introduction of macromolecules into cells. BioTechniques. 1988;**6**:742-751

[15] Sherba JJ, Hogquist S, Lin H, Shan JW, Shreiber DI, Zahn JD. The effects of electroporation buffer composition on cell viability and electro-transfection efficiency. Scientific Reports. 2020;**10**:3053

[16] Li S. Optimizing electrotransfection of Mammalian cells in vitro. Cold Spring Harbor Protocols. 2006;**1**:pdb-prot4449

[17] Flanagan M, Gimble JM, Yu G, Wu X, Li S. Competitive electroporation formulation for cell therapy. Cancer Gene Therapy. 2011;**18**:579-586

[18] Ye X, Brown SK, Scorza R, Cordts J, Sanford JC. Genetic transformation of peach tissues by particle bombardment. Journal of the American Society for Horticultural Science. 1994;**119**:367-373

[19] Uchida M, Li XW, Mertens P, Alpar HO. Transfection by particle bombardment: Delivery of plasmid DNA into mammalian cells using gene gun. Biochimica et Biophysica Acta (BBA) - General Subjects. 2009;**1790**:754-764

[20] Svarovsky SA, Gonzalez-Moa MJ, Robida MD, Borovkov AY, Sykes K. Self-assembled micronanoplexes for improved biolistic delivery of nucleic acids. Molecular Pharmaceutics. 2009;**6**:1927-1933

[21] Tsukakoshi M, Kurata S, Nomiya Y, Ikawa Y, Kasuya T. A novel method of DNA transfection by laser microbeam cell surgery. Applied Physics B: Lasers and Optics. 1984;**35**:135-140

[22] Rhodes K, Clark I, Zatcoff M, Eustaquio T, Hoyte KL, Koller MR. Cellular laserfection. Methods in Cell Biology. 2007;**82**:309-333

[23] Krasieva TB, Chapman CF, LaMorte VJ, Venugopalan V, Berns MW, Tromberg BJ. Cell permeabilization and molecular transport by laser microirradiation. In: Optical Investigations of Cells In Vitro and In Vivo. Bellingham, Washington, United States: SPIE; 1998. pp. 38-44

[24] Pislaru SV, Greenleaf JF, Miller DL. Sonoporation: Mechanical DNA Delivery by Ultrasonic Cavitation. Kluwer Academic Publishers-Plenum Publishers; Plenum Publishing Corporation ; Springer Science+Business Media; 2002

[25] Mignet N, Marie C, Delalande A, Manta S, Bureau M-F, Renault G, et al. Microbubbles for nucleic acid delivery in liver using mild Sonoporation. Methods in Molecular Biology. 1943;**2019**: 377-387

[26] Zhang Z, Wang Y, Zhang H, Tang Z, Liu W, Lu Y, et al. Hypersonic Poration: A new versatile cell Poration method to enhance cellular uptake using a piezoelectric Nano-electromechanical device. Small. 2017;**13**

[27] Sharei A, Zoldan J, Adamo A, Sim WY, Cho N, Jackson E, et al. A vector-free microfluidic platform for intracellular delivery. PNAS. 2013;**110**:2082-2087

[28] Chakrabarty P, Gupta P, Illath K, Kar S, Nagai M, Tseng F-G, et al. Microfluidic mechanoporation for cellular delivery and analysis. Materials Today Bio. 2022;**13**:100193

[29] Huth S, Lausier J, Gersting SW, Rudolph C, Plank C, Welsch U, et al. Insights into the mechanism of magnetofection using PEI-based magnetofectins for gene transfer. The Journal of Gene Medicine. 2004;**6**:923-936

[30] Scherer F, Anton M, Schillinger U, Henke J, Bergemann C, Krüger A, et al. Magnetofection: Enhancing and targeting gene delivery by magnetic force in vitro and in vivo. Gene Therapy. 2002;**9**:102-109

[31] Kami D, Takeda S, Itakura Y, Gojo S, Watanabe M, Toyoda M. Application of magnetic nanoparticles to gene delivery. IJMS. 2011;**12**:3705-3722

[32] Kim TK, Eberwine JH. Mammalian cell transfection: The present and the future. Analytical and Bioanalytical Chemistry. 2010;**397**:3173-3178

[33] Graham FL, van der Eb AJ. A new technique for the assay of infectivity

of human adenovirus 5 DNA. Virology. 1973;**52**:456-467

[34] Guo L, Wang L, Yang R, Feng R, Li Z, Zhou X, et al. Optimizing conditions for calcium phosphate mediated transient transfection. Saudi Journal of Biological Sciences. 2017;**24**:622-629

[35] Batard P, Jordan M, Wurm F. Transfer of high copy number plasmid into mammalian cells by calcium phosphate transfection. Gene. 2001;**270**:61-68

[36] Pick HM, Meissner P, Preuss AK, Tromba P, Vogel H, Wurm FM. Balancing GFP reporter plasmid quantity in large-scale transient transfections for recombinant anti-human rhesus-D IgG1 synthesis. Biotechnology and Bioengineering. 2002;**79**:595-601

[37] Schenborn ET, Goiffon V. Calcium phosphate transfection of mammalian cultured cells. In: Tymms MJ, editor. Transcription Factor Protocols [internet]. Totowa, NJ: Humana Press; 2000. p. 135-145 10.1385/1-59259-686-X:135

[38] Haberland A, Knaus T, Zaitsev SV, Stahn R, Mistry AR, Coutelle C, et al. Calcium ions as efficient cofactor of polycation-mediated gene transfer. Biochimica et Biophysica Acta (BBA) - Gene Structure and Expression. 1999;**1445**:21-30

[39] Lindell J, Girard P, Müller N, Jordan M, Wurm F. Calfection: A novel gene transfer method for suspension cells. Biochimica et Biophysica Acta. 2004;**1676**(2):155-161

[40] Sussman DJ, Milman G. Short-term, high-efficiency expression of transfected DNA. Molecular and Cellular Biology. 1984;**4**8:1641-1643

[41] Kumar P, Nagarajan A, Uchil PD. DEAE-Dextran Transfection. Cold Spring Harbor, New York, United States: Cold Spring Harbor Laboratory Press; 2018

[42] Chesnoy S, Huang L. Structure and function of lipid-DNA complexes for gene delivery. Annual Review of Biophysics and Biomolecular Structure. 2000;**29**:27-47

[43] Kumar P, Nagarajan A, Uchil PD. Lipofection. Cold Spring Harbor, New York, United States: Cold Spring Harbor Laboratory Press; 2019

[44] Felgner PL, Gadek TR, Holm M, Roman R, Chan HW, Wenz M, et al. Lipofection: A highly efficient, lipid-mediated DNA-transfection procedure. Proceedings of the National Academy of Sciences of the United States of America. 1987;**84**:7413-7417

[45] El Ouahabi A, Thiry M, Pector V, Fuks R, Ruysschaert JM, Vandenbranden M. The role of endosome destabilizing activity in the gene transfer process mediated by cationic lipids. FEBS Letters. 1997;**414**:187-192

[46] Stürzl M, Konrad A, Sander G, Wies E, Neipel F, Naschberger E, et al. High throughput screening of gene functions in mammalian cells using reversely transfected cell arrays: Review and protocol. Combinatorial Chemistry & High Throughput Screening. 2008;**11**:159-172

[47] Dalby B, Cates S, Harris A, Ohki EC, Tilkins ML, Price PJ, et al. Advanced transfection with lipofectamine 2000 reagent: Primary neurons, siRNA, and high-throughput applications. Methods. 2004;**33**:95-103

[48] Boussif O, Lezoualc'h F, Zanta MA, Mergny MD, Scherman D, Demeneix B, et al. A versatile vector for gene and oligonucleotide transfer into cells in

culture and in vivo: Polyethylenimine. PNAS. 1995;**92**:7297-7301

[49] Merdan T, Kunath K, Fischer D, Kopecek J, Kissel T. Intracellular processing of poly(ethylene imine)/ ribozyme complexes can be observed in living cells by using confocal laser scanning microscopy and inhibitor experiments. Pharmaceutical Research. 2002;**19**:140-146

[50] Sonawane ND, Szoka FC, Verkman AS. Chloride accumulation and swelling in endosomes enhances DNA transfer by polyamine-DNA polyplexes*. Journal of Biological Chemistry. 2003;**278**:44826-44831

[51] Gilboa E. Retrovirus vectors and their uses in molecular biology. BioEssays. 1986;**5**:252-257

[52] Finkelshtein D, Werman A, Novick D, Barak S, Rubinstein M. LDL receptor and its family members serve as the cellular receptors for vesicular stomatitis virus. Proceedings of the National Academy of Sciences of the United States of America. 2013;**110**:7306-7311

[53] Yamauchi Y, Greber UF. Principles of virus uncoating: Cues and the snooker ball. Traffic. 2016;**17**:569-592

[54] Miyanohara A. Preparation of vesicular stomatitis virus-G (VSV-G) conjugate and its use in gene transfer. Cold Spring Harbor Protocols. 2012;**2012**:453-456

[55] Mangion M, Robert M-A, Slivac I, Gilbert R, Gaillet B. Production and use of Gesicles for nucleic acid delivery. Molecular Biotechnology. 2022;**64**:278-292

[56] Balistreri G, Yamauchi Y, Teesalu T. A widespread viral entry mechanism: The C-end Rule motif–neuropilin receptor interaction. PNAS. 2021;**118**(49):e2112457118

[57] Wadia JS, Dowdy SF. Transmembrane delivery of protein and peptide drugs by TAT-mediated transduction in the treatment of cancer. Advanced Drug Delivery Reviews. 2005;**57**:579-596

[58] Dupont E, Prochiantz A, Joliot A. Penetratin story: An overview. In: Langel Ü, editor. Cell-Penetrating Peptides: Methods and Protocols. New York, NY: Springer; 2015. pp. 29-37. DOI: 10.1007/978-1-4939-2806-4_2

[59] Derakhshankhah H, Jafari S. Cell penetrating peptides: A concise review with emphasis on biomedical applications. Biomedicine & Pharmacotherapy. 2018;**108**: 1090-1096

[60] Ignatovich IA, Dizhe EB, Pavlotskaya AV, Akifiev BN, Burov SV, Orlov SV, et al. Complexes of plasmid DNA with basic domain 47-57 of the HIV-1 tat protein are transferred to mammalian cells by endocytosis-mediated pathways*. Journal of Biological Chemistry. 2003;**278**: 42625-42636

[61] EL Andaloussi S, Lehto T, Mäger I, Rosenthal-Aizman K, Oprea II, Simonson OE, et al. Design of a peptide-based vector, PepFect6, for efficient delivery of siRNA in cell culture and systemically in vivo. Nucleic Acids Research. 2011;**39**:3972-3987

[62] Ezzat K, EL Andaloussi S, Zaghloul EM, Lehto T, Lindberg S, Moreno PMD, et al. PepFect 14, a novel cell-penetrating peptide for oligonucleotide delivery in solution and as solid formulation. Nucleic Acids Research. 2011;**39**:5284-5298

[63] Veiman K-L, Mäger I, Ezzat K, Margus H, Lehto T, Langel K, et al. PepFect14 peptide vector for efficient gene delivery in cell cultures. Molecular Pharmaceutics. 2013;**10**:199-210

[64] Ronquist G, Hedström M. Restoration of detergent-inactivated adenosine triphosphatase activity of human prostatic fluid with concanavalin A. Biochimica et Biophysica Acta (BBA) - Enzymology. 1977;**483**:483-486

[65] Waldenström A, Ronquist G. Role of exosomes in myocardial remodeling. Circulation Research. 2014;**114**:315-324

[66] Nilsson BO, Lennartsson L, Carlsson L, Nilsson S, Ronquist G. Expression of prostasome-like granules by the prostate cancer cell lines PC3, Du145 and LnCaP grown in monolayer. Upsala Journal of Medical Sciences 1999;**104**(3):199-206

[67] Munagala R, Aqil F, Jeyabalan J, Kandimalla R, Wallen M, Tyagi N, et al. Exosome-mediated delivery of RNA and DNA for gene therapy. Cancer Letters. 2021;**505**:58-72

[68] Inoue N, Hirouchi T, Kasai A, Higashi S, Hiraki N, Tanaka S, et al. Unbiased compound screening with a reporter gene assay highlights the role of p13 in the cardiac cellular stress response. Biochemical and Biophysical Research Communications. 2018;**495**:1992-1997

[69] Arena TA, Harms PD, Wong AW. High throughput transfection of HEK293 cells for transient protein production. Methods in Molecular Biology. 1850;**2018**:179-187

[70] Fiscella M, Perry JW, Teng B, Bloom M, Zhang C, Leung K, et al. TIP, a T-cell factor identified using high-throughput screening increases survival in a graft-versus-host disease model. Nature Biotechnology. 2003;**21**:302-307

[71] Madoux F, Simanski S, Chase P, Mishra JK, Roush WR, Ayad NG, et al. An ultra-high throughput cell-based screen for Wee1 degradation inhibitors. The Journal of Biomolecular Screening. 2010;**15**:907-917

[72] Buchser WJ, Pardinas JR, Shi Y, Bixby JL, Lemmon VP. 96-well electroporation method for transfection of mammalian central neurons. BioTechniques. 2006;**41**:619-624

[73] Guignet EG, Meyer T. Suspended-drop electroporation for high-throughput delivery of biomolecules into cells. Nature Methods. 2008;**5**:393-395

[74] Martinet W, Schrijvers DM, Kockx MM. Nucleofection as an efficient nonviral transfection method for human monocytic cells. Biotechnology Letters. 2003;**25**:1025-1029

[75] Trompeter H-I, Weinhold S, Thiel C, Wernet P, Uhrberg M. Rapid and highly efficient gene transfer into natural killer cells by nucleofection. Journal of Immunological Methods. 2003;**274**:245-256

[76] Mellott AJ, Forrest ML, Detamore MS. Physical non-viral gene delivery methods for tissue engineering. Annals of Biomedical Engineering. 2013;**41**:446-468

[77] SF Cell Line 96-well Nucleofector™ Kit (96 RCT)|Lonza. 2022. Available from: https://bioscience.lonza.com/lonza_bs/FR/en/Transfection/p/000000000000198709/SF-Cell-Line-96-well-Nucleofector-Kit

[78] Haag J, Voigt R, Soeder S, Aigner T. Efficient non-viral transfection of primary human adult chondrocytes in a high-throughput format. Osteoarthritis and Cartilage. 2009;**17**:813-817

[79] Bowles R, Patil S, Pincas H, Sealfon SC. Validation of efficient high-throughput plasmid and siRNA transfection of human monocyte-derived dendritic cells without cell maturation. Journal of Immunological Methods. 2010;**363**:21-28

[80] Moore JC, Atze K, Yeung PL, Toro-Ramos AJ, Camarillo C, Thompson K, et al. Efficient, high-throughput transfection of human embryonic stem cells. Stem Cell Research & Therapy. 2010;**1**:23

[81] 384-well Nucleofector™ System|Lonza. 2022. Available from: https://bioscience.lonza.com/lonza_bs/FR/en/Transfection/p/000000000000198370/384-well-Nucleofector-System

[82] Electrode, Petri Pulser™. VWR. 2022. Available from: https://ru.vwr.com/store/product/10597200/electrode-petri-pulsertm

[83] Petri Dish Electrode. VWR. 2022 Available from: https://ru.vwr.com/store/product/10597253/petri-dish-electrode

[84] BTX Adherent Cell Electrodes - Clinical Analyzers and Instruments, Electroporators. 2022. Available from: https://www.fishersci.com/shop/products/adherent-cell-electrodes-2/p-7222369

[85] Pihl J, Johansson M-L, Granfeldt D, Tokarz M, Karlsson M, Sinclair J. Cellaxess®HT: high-throughput transfection for genome-wide RNAi. Nature Methods. 2008;**5**:i-ii

[86] Marine S, Freeman J, Riccio A, Axenborg M-L, Pihl J, Ketteler R, et al. High-throughput transfection of differentiated primary neurons from rat forebrain. Journal of Biomolecular Screening. 2012;**17**:692-696

[87] Fujimoto H, Kato K, Iwata H. Electroporation microarray for parallel transfer of small interfering RNA into mammalian cells. Analytical and Bioanalytical Chemistry. 2008;**392**:1309-1316

[88] Jain T, Papas A, Jadhav A, McBride R, Saez E. In situ electroporation of surface-bound siRNAs in microwell arrays lab chip. The Royal Society of Chemistry. 2012;**12**:939-947

[89] Bian S, Zhou Y, Hu Y, Cheng J, Chen X, Xu Y, et al. High-throughput in situ cell electroporation microsystem for parallel delivery of single guide RNAs into mammalian cells. Scientific Reports. 2017;**7**:42512

[90] Zhang P, Zhang J, Bian S, Chen Z, Hu Y, Hu R, et al. High-throughput superhydrophobic microwell arrays for investigating multifactorial stem cell niches. Lab on a Chip. 2016;**16**:2996-3006

[91] Yamauchi F, Kato K, Iwata H. Spatially and temporally controlled gene transfer by electroporation into adherent cells on plasmid DNA-loaded electrodes. Nucleic Acids Research. 2004;**32**:e187

[92] Fox MB, Esveld DC, Valero A, Luttge R, Mastwijk HC, Bartels PV, et al. Electroporation of cells in microfluidic devices: A review. Analytical and Bioanalytical Chemistry. 2006;**385**:474-485

[93] Huang Y, Rubinsky B. Flow-through micro-electroporation chip for high efficiency single-cell genetic manipulation. Sensors and Actuators A: Physical. 2003;**104**(3):205-212

[94] Hur J, Chung AJ. Microfluidic and nanofluidic intracellular delivery. Advanced Science. 2021;**8**:2004595

[95] Kim JA, Cho K, Shin MS, Lee WG, Jung N, Chung C, et al. A novel electroporation method using a capillary and wire-type electrode. Biosensors & Bioelectronics. 2008;**23**:1353-1360

[96] Li Y, Wu M, Zhao D, Wei Z, Zhong W, Wang X, et al. Electroporation on microchips: The harmful effects of pH changes and scaling down. Scientific Reports. 2015;**5**:17817

[97] He H, Chang DC, Lee Y-K. Using a micro electroporation chip to determine the optimal physical parameters in the uptake of biomolecules in HeLa cells. Bioelectrochemistry. 2007;**70**:363-368

[98] Wang S, Zhang X, Wang W, Lee LJ. Semicontinuous flow electroporation chip for high-throughput transfection on mammalian cells. Analytical Chemistry. 2009;**81**:4414-4421

[99] Kang W, Giraldo-Vela JP, NathamgariSSP,McGuireT,McNaughtonRL, Kessler JA, et al. Microfluidic device for stem cell differentiation and localized electroporation of postmitotic neurons. Lab on a Chip. 2014;**14**:4486-4495

[100] Nasoff M, Bergseid M, Hoeffler JP, Heyman JA. High-throughput expression of fusion proteins. Methods in Enzymology. 2000;**328**:515-529

[101] Colin B, Deprez B, Couturier C. High-throughput DNA plasmid transfection using acoustic droplet ejection technology. In: SLAS DISCOVERY: Advancing the Science of Drug Discovery. Vol. 24. Thousand Oaks, CA: SAGE Publications Inc STM; 2019. pp. 492-500

[102] Sichero L, Sobrinho JS, Villa LL. Identification of novel cellular transcription factors that regulate early promoters of human papillomavirus types 18 and 16. The Journal of Infectious Diseases. 2012;**206**:867-874

[103] Lievens S, Vanderroost N, Van der Heyden J, Gesellchen V, Vidal M, Tavernier J. Array MAPPIT: High-throughput interactome analysis in mammalian cells. Journal of Proteome Research. 2009;**8**:877-886

[104] Yokoyama S, Ito Y, Ueno-Kudoh H, Shimizu H, Uchibe K, Albini S, et al. A systems approach reveals that the myogenesis genome network is regulated by the transcriptional repressor RP58. Developmental Cell. 2009;**17**:836-848

[105] Colin B, Rocq N, Deprez B, Couturier C. High-throughput DNA plasmid multiplexing and transfection using acoustic nanodispensing technology. JoVE (Journal of Visualized Experiments). 2019;**150**:e59570

[106] Ziauddin J, Sabatini DM. Microarrays of cells expressing defined cDNAs. Nature. 2001;**411**:107-110

[107] Palmer E, Freeman T. Investigation into the use of C- and N-terminal GFP fusion proteins for subcellular localization studies using reverse transfection microarrays. Comparative and Functional Genomics. 2004;**5**:342-353

[108] Delehanty JB, Shaffer KM, Lin B. Transfected cell microarrays for the expression of membrane-displayed single-chain antibodies. Analytical Chemistry. 2004;**76**:7323-7328

[109] Palmer EL, Miller AD, Freeman TC. Identification and characterisation of human apoptosis inducing proteins using cell-based transfection microarrays and expression analysis. BMC Genomics. 2006;**7**:145

[110] Konrad A, Jochmann R, Kuhn E, Naschberger E, Chudasama P, Stürzl M. Reverse transfected cell microarrays in

infectious disease research. Methods in Molecular Biology. 2011;**706**:107-118

[111] Walczak W, Pipalia NH, Soni M, Faruqi AF, Ralph H, Maxfield FR, et al. Parallel analysis of v-Src mutant protein function using reverse transfection cell arrays. Combinatorial Chemistry & High Throughput Screening;**9**:711-718

[112] Rajan S, Djambazian H, Dang HCP, Sladek R, Hudson TJ. The living microarray: A high-throughput platform for measuring transcription dynamics in single cells. BMC Genomics. 2011;**12**:115

[113] Lievens S, Van der Heyden J, Masschaele D, De Ceuninck L, Petta I, Gupta S, et al. Proteome-scale binary Interactomics in human cells. Molecular & Cellular Proteomics. 2016;**15**:3624-3639

[114] Schudel BR, Harmon B, Abhyankar VV, Pruitt BW, Negrete OA, Singh AK. Microfluidic platforms for RNA interference screening of virus–host interactions. Lab Chip. 2013;**13**:811-817

[115] Enomoto J, Takagi R, Onuki-Nagasaki R, Fujita S, Fukuda J. Reverse transfection in microchamber arrays for cell migration assays. Sensors and Actuators B: Chemical. 2014;**190**:896-899

[116] Woodruff K, Maerkl S. A high-throughput microfluidic platform for mammalian cell transfection and culturing. Scientific Reports. 2016;**6**:23937

[117] Škalamera D, Ranall MV, Wilson BM, Leo P, Purdon AS, Hyde C, et al. A High-throughput platform for lentiviral overexpression screening of the human ORFeome. PLOS ONE, Public Library of Science. 2011;**6**:e20057

[118] Yang X, Boehm JS, Yang X, Salehi-Ashtiani K, Hao T, Shen Y, et al. A public genome-scale lentiviral expression library of human ORFs. Nature Methods. 2011;**8**:659-661

[119] Zhang J, Hu Y, Wang X, Liu P, Chen X. High-throughput platform for efficient chemical transfection, virus packaging, and transduction. Micromachines. 2019;**10**:387

[120] Streamlined Production of High-titer Lenti Virus in a 96-well format. 2022. Available from: https://www.takarabio.com/learning-centers/gene-function/viral-transduction/lentivirus/high-throughput-lentivirus-production

[121] Ochiya T, Takahama Y, Nagahara S, Sumita Y, Hisada A, Itoh H, et al. New delivery system for plasmid DNA in vivo using atelocollagen as a carrier material: The Minipellet. Nature Medicine. 1999;**5**:707-710

[122] Honma K, Ochiya T, Nagahara S, Sano A, Yamamoto H, Hirai K, et al. Atelocollagen-based gene transfer in cells allows high-throughput screening of gene functions. Biochemical and Biophysical Research Communications. 2001;**289**:1075-1081

[123] Saito S, Honma K, Kita-Matsuo H, Ochiya T, Kato K. Gene expression profiling of cerebellar development with high-throughput functional analysis. Physiological Genomics. 2005;**22**:8-13

[124] Redmond TM, Ren X, Kubish G, Atkins S, Low S, Uhler MD. Microarray transfection analysis of transcriptional regulation by cAMP-dependent protein kinase*. Molecular & Cellular Proteomics. 2004;**3**:770-779

[125] Gautam A, Singh H, Tyagi A, Chaudhary K, Kumar R, Kapoor P, et al.

CPPsite: A curated database of cell penetrating peptides. Database (Oxford). 2012;**2012**:bas015

[126] Agrawal P, Bhalla S, Usmani SS, Singh S, Chaudhary K, Raghava GPS, et al. CPPsite 2.0: A repository of experimentally validated cell-penetrating peptides. Nucleic Acids Research. 2016;**44**:D1098-D1103

[127] de Figueiredo IR, Freire JM, Flores L, Veiga AS, Castanho MARB. Cell-penetrating peptides: A tool for effective delivery in gene-targeted therapies. IUBMB Life. 2014;**66**:182-194

[128] Cunningham F, Amode MR, Barrell D, Beal K, Billis K, Brent S, et al. Ensembl 2015. Nucleic Acids Research. 2015;**43**:D662-D669

[129] Alam MK, Koomson E, Zou H, Yi C, Li C-W, Xu T, et al. Recent advances in microfluidic technology for manipulation and analysis of biological cells (2007-2017). Analytica Chimica Acta. 2018;**1044**:29-65

Chapter 2

Genetic Transformation in Prokaryotic and Eukaryotic Cells

Endang Semiarti, Yekti Asih Purwestri, Saifur Rohman and Wahyu Aristyaning Putri

Abstract

Improving the quality and quantity of an organism and its products can be approached by molecular characters enhancement through the insertion of a gene of interest into cells of the desired organism. Genetic transformation of an organism involves isolation, identification, cloning a gene of interest into a vector, and transferring the gene to the target organism. This chapter reviews the process of genetic transformation into the organism's cell from bacterial (*Escherichia coli*), yeast, plant (Onion, Tobacco, and Orchids), and mammalian. The discussion will be focused on the introduction of DNA molecules into plant cells and protoplast mediated by polyethylene glycol (PEG), electroporation, and gene gun using particle bombardment. Further discussion on the transient protein expression system of plant-based on protoplast, onion cell, and tobacco will also be covered in this chapter as well. The systems have been proven as a powerful tool for determining subcellular protein localization, protein-protein interactions, identifying gene function, and regulation. Finally, it can be clearly seen, the differences and similarities in the mechanism of genetic transformation both in prokaryotic and eukaryotic systems.

Keywords: eukaryotic cells, genetic transformation, molecular character enhancement, prokaryotic cells, transient expression

1. Introduction

To improve the quality and quantity of an organism, both prokaryotes and eukaryotes, it can be approached by molecular character enhancement through the insertion of interest genes or superior genes into the cells of the desired organism. The process of genetic transformation of an organism involves the isolation and identification of the gene of interest, the technique of cloning the gene on a plasmid vector until the process of transferring the gene to the target organism's cell. One of the important genes in the growth of organisms is the homeobox gene, which is a gene that regulates the growth and development of organisms in a very early stage. Homeobox genes were first discovered in the *Drosophila melanogaster*. These homeobox genes have been also found in all multicellular organisms from fungi to plants, and vertebrate animals [1].

In plants, overexpression of the homeobox gene at an early stage of growth will activate the formation of apical buds from apical meristems that will produce shoots.

The addition of exogenous cytokinin and auxin growth regulators will activate the homeobox genes to induce cell division genes that in turn will produce somatic embryos. Theoretically, each somatic cell can grow and transform itself into somatic

Gene	Function	Organism	Ref
OSH1	Homologous with Kn1 (*Zea mays*). Altered morphology of transgenic plants	*Oryza sativa*	[3, 4]
OSH15	Homologous with Kn1 (Zea mays). Affect the design of internodes resulting in stunted plants	*Oryza sativa*	[5]
KNAT1 and *KNAT2*	Morphogenesis	*Arabidopsis thaliana*	[6]
blr-1 and *blr-2*	Phyllotactic pattern and stem cell fate	*Arabidopsis thaliana*	[7]
ANL2	Accumulation of anthocyanin and in root development	*Arabidopsis thaliana*	[8]
ATHB-2	Mediates plant morphogenesis for light signals response	*Arabidopsis thaliana*	[9]
MDH1	Homologous *BEL1* (*Arabidopsis*). Play an essential role in the control of plant fertility	*Malus domestica* Borkh	[10]
ATH1	Controls plant architecture by locally restricting environmental responses	*Arabidopsis thaliana*	[11]
GLABRA2	Effects seed oil content in *Arabidopsis*	*Arabidopsis thaliana*	[12]
WOX9	have species-specific roles in embryo and inflorescence development	*Arabidopsis thaliana*	[13]
WOX4	Regular *TDIF* Peptide Signaling Regulates Vascular Stem Cell Proliferation	*Arabidopsis thaliana*	[14]
LeT6	The leaves morphological states	Tomato	[15]
Athb-12	Response to treatment for abscisic acid (ABA)	*Arabidopsis thaliana*	[16]
KNAT6	Involved in Meristem Activity and Organ Separation	*Arabidopsis thaliana*	[17]
ATHB16	Regulates the photoperiod sensitivity in *Arabidopsis* and leaf development	*Arabidopsis thaliana*	[18]
ATK1	Transcriptional activator	*Arabidopsis thaliana*	[19]
ZmOCL1	Play a role in the embryo protoderm identity specification, organize of the primary root primordium or the L1 cell layer maintenance in the shoot apical meristem	*Zea mays*	[20]
BELL1	Master regulator for the gametophyte-to-sporophyte transition	*Physcomitrella patens*	[21]
Athb-2	Changes in light quality perceived by a novel phytochrome regulate plant development	Green plants	[22]
NTH15	Regulate leaf and flower morphology, accompanied by a decrease in the content of the active gibberellin		[14]

Table 1.
The homeobox gene in plants.

embryos, therefore it can produce plant seeds in large quantities and uniform phenotypic characters. This is very profitable for agriculture and industry, especially for the mass production of identical plant seeds using tissue culture techniques.

In the model plant, *Arabidopsis thaliana,* it has been reported that the homeobox genes always maintain the growth of meristem cells in Shoot Apical Meristem (SAM) [2]. Overexpression of the homeobox gene in Arabidopsis has shown that the cells can convert from a determinate state to the meristematic indeterminate state, depending on the levels of expression of the gene (s) (**Table 1**) [23].

2. Transformation for transient expression in onion, tobacco leaves, and protoplast

Transient expression become a powerful tool in functional genomics study for detecting gene expression in a short time and the inserted gene do not integrate into the plant genome. A transient expression system has been developed in planta using different cells or tissues, including protoplast, onion cells, and tobacco (*Nicotiana benthamiana*) leaves (**Table 2**). A transient expression system using protoplasts has proven to be a good experimental tool in molecular biology. This approach is an efficient technique to study subcellular protein localization, protein complexes, *in vivo* gene silencing, and promotor activity [24, 25].

The advantages of the transient expression system compared to stable expression are that it does not require regeneration of transformed cells, does not affect the stability of the host genome, and is independent of the effect of T-DNA integration site position [28]. Protoplast transfection can be performed using a variety of procedures commonly used for the transfection of animal cell cultures. The procedures that are often used to insert DNA into protoplasts are polyethylene glycol (PEG) and electroporation [29].

Polyethylene glycol (PEG)-mediated transformation plant cells can be transformed through certain chemicals, namely PEG (polyethylene glycol). PEG is an oligomer or hydrophilic polymer synthesized from ethylene oxide, containing repeating units of -(O-CH_2-CH_2)-. Polyethylene oxide (PEO) is another name for PEG. Typically, ethylene oxide macromolecules with a molecular weight of less than 20,000 g/mol are called PEGs, while macromolecules with values above 20,000 g/molar are called PEOs [29]. PEG is soluble in acetonitrile, benzene, water, ethanol, and dichloromethane, while it is insoluble in diethyl ether and hexane (**Figures 1** and **2**).

PEG is available in various structures, such as branched, stellar, and comb-like macromolecules. PEG can bind various reactive functional groups to the PEG polymer site. Homo and heterobifunctional PEG derivatives are particularly suitable as agents

Plant materials	Transformation methods	Purposes/Functions	Refs.
Protoplast	PEG	Subcellular localization of proteins	[24, 25]
Nicotiana benthamiana leaves	Agroinfiltration	Metabolite's production (protein, secondary metabolite, etc.)	[26]
Onion cells	Particle bombardment	Subcellular localization of proteins	[27]

Table 2.
Transient expression system and its purposes in planta.

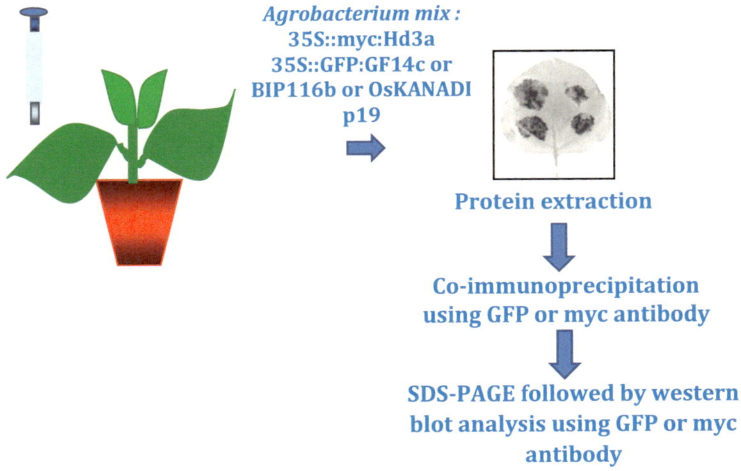

Figure 1.
Agroinfiltration in tobacco (Nicotiana benthamiana) leaves for protein-protein analysis.

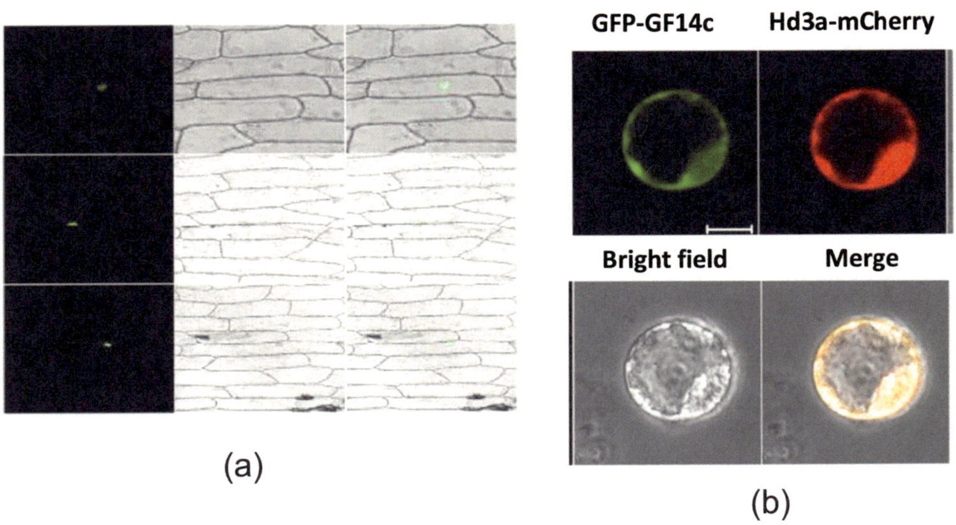

Figure 2.
Transient expression in onion cell and protoplast for determining the subcellular localization of the protein. (a) Subcellular localization of OsKAN1-GFP fusion protein in the nucleus of onion cell transformed using particle bombardment [30]. (b) Transient expression of GFP-GF14c and Hd3a-mCherry in rice protoplast was driven by the 35S promoter of cauliflower mosaic virus and ubiquitin promoter, respectively, Bar = 10 μm.

or spacers of two chemical entities, whereas mono-functional PEGs prevent linking reactions that can affect the PEGylation of certain compounds with bifunctional PEGs. PEGylation is an interesting process in which PEG is bound to other molecules [31, 32].

PEG was used to increase DNA uptake into the protoplast during transfection. Very high concentrations of PEG can reduce transfection efficiency because it is toxic to protoplasts [33]. PEG-mediated DNA uptake is a direct gene transfer method that utilizes the interaction between PEG, naked DNA, salts, and protoplast membranes to influence the transport of DNA into the cytoplasm. The advantage of PEG-mediated transformation is that it does not require special equipment and can be carried out in the laboratory under sterile conditions [34]. Compared to *Agrobacterium tumefaciens*-mediated transformation, PEG-mediated transformation was not species-specific. In addition, PEG-mediated transformation is also useful for functional analysis of genes through transient expression, a technique that is often used for promoter analysis [35].

Particle bombardment particles are coated in DNA and can penetrate plant cells without killing the plant cells themselves. Previous experiments have shown that particle bombardment has been successfully used to insert DNA into rice callus and seedlings grown in dark conditions but has the disadvantage of low efficiency and reliance on expensive equipment [36].

3. Expression of recombinant psychrophilic RNase III in *Escherichia coli*

To understand the mechanism of how the transformation and expression of recombinant protein in a prokaryotic system, *Escherichia coli* BL21(DE3) have been used as host and recombinant RNaseIII as a model protein. Ribonuclease III is an enzyme that specifically cleaves the double-stranded RNA molecules. It functions for ribosomal RNA maturation; therefore, RNase III is indispensable for the survival of cells. Here, the production of recombinant psychrophilic RNase III from *Shewanella* sp. SIB1 in the *Escherichia coli* system was reported. As a psychrophilic enzyme, recombinant RNase III was produced in the form of inclusion bodies. To produce the soluble recombinant psychrophilic RNase III, co-expression with FKBP22 from the same bacteria was carried out. The result showed that FKBP22 significantly improved the solubility of recombinant psychrophilic RNase III. It strongly suggested that FKBP22 assists the proper folding of recombinant psychrophilic RNase III when it was overproduced in the *Escherichia coli* system.

Ribonuclease III (RNase III) is an enzyme that specifically cleaves double-stranded RNA [30, 37–40]. RNase III has an important role in both the RNA transcript maturation and decay of diverse cellular and viral RNA. A primary function of RNase III, however, is the maturation of ribosomal RNA (rRNA) [30, 37, 38, 40, 41]. RNase III has been known to be widely distributed across the living kingdom of life, from bacteria to higher eukaryotes. RNase III family has common features in their molecular organization, by which it consists of catalytic domain with the common feature of HNERLFGDS located at the N-terminus and double-stranded binding domain (dsRBD) that located on their C-terminus [39]. RNase III exhibited enzymatically active in homodimeric form, by which each monomer has its catalytic mechanism and therefore the cleavage product of the RNase III exhibits a very regular length of short double-stranded RNA [39]. By such properties, RNase III can be manipulated to produce short dsRNA that can be implemented for the RNA interference technology in combination with Argonaute, Drosha, and Dicer [42]. Therefore, the production of recombinant RNase III is necessary from the scientific and technological point of view.

Production of recombinant proteins could be done in either bacterial or mammalian cells as a host. The choice of the host to produce recombinant protein may be the subject of proteins of interest. It depends on whether further processing of the proteins of interest is necessary or not. However, the bacterial cell is the most prominent host for recombinant protein production. *Escherichia coli* is the most common bacterial cell that is generally used as a host organism because of the following advantages—(a) it has unparalleled fast growth kinetics, (b) high cell density cultures are easily achieved, (c) the growth media are easily prepared and inexpensive, and (d) transformation with exogenous DNA is fast and easy [43]. There are several commercially available *Escherichia coli* appropriate for the expression host of recombinant proteins, such as *Escherichia coli* BL21(DE3) and its derivatives. *Escherichia coli* BL21(DE3) is carrying the T7gene1 from the lysogens DE3, a derivative of bacteriophage lambda, that encodes for T7 RNA polymerase under the control of *lac*UV5 promoter [44]. T7 RNA polymerase is a polymerase that can recognize T7 promoter, a strong promoter appropriate for the high-level expression of proteins. Such promoter is commonly used in several commercially available expression vectors, such as pET series, pRSET, and pACYC-Duet. These vectors contain a regulatory system in the form of *lac*I in which the gene product suppresses the expression of recombinant proteins.

This report will discuss the production of recombinant RNase III from a psychrotrophic bacterium, *Shewanella* sp. SIB1. *Shewanella* sp. SIB1 is a psychrotrophic bacterium that grows most rapidly at 20°C [45]. This strain can grow even at 0° but cannot grow higher than 30°C. Phylogenetic analysis indicates that *Shewanella* sp. SIB1 is closely related to the *Shewanella* sp. AC10 isolated from the Antarctic ocean [44]. Interestingly, protein from psychrotrophic bacterium exhibits distinct properties compared to the mesophilic counterparts by their ability to adapt to cold temperatures [45].

Protein adaptation in such low temperatures requires a strategy that is not commonly found in mesophilic, for example, psychrophilic proteins must be flexible enough to avoid the problem in protein folding and to perform the optimum catalytic activity if it is an enzyme. Therefore, the production of psychrophilic protein would be interesting due to their properties to adapt to such low temperatures. Although the production of recombinant protein in bacterial host seems to be straightforward, several difficulties that arise and how to solve the problems during the production of recombinant psychrophilic protein will be discussed.

3.1 Localization of Shewanella sp. SIB1 RNase III encoding gene (*Sh-rnc*)

To localize the *Sh-rnc* gene from the *Shewanella* sp. SIB1 genome, as well as to obtain the full length of the RNase III open reading frame, the inverse PCR was carried out in this work. Previously, the partial *Sh-rnc* gene was amplified by using a pair of primers constructed based on the sequence of open reading frames of the *rnc* gene from *Shewanella oneidensis* MR1. Once the fragment of the *Sh-rnc* gene was amplified then it was used to construct new primers for the inverse PCR. For the inverse PCR, the SIB1 genome was digested by the *Dra*I restriction enzyme and then the digestion product was then allowed to perform self-ligation to form small circular products. Since the *orf* of the *Sh-rnc* gene contains a recognition site for *Dra*I, therefore, the PCR was conducted by using two pairs of primers. By such a strategy, the two PCR products were obtained and were then cloned into pUC18 for sequencing. The sequencing results indicated that the two fragments corresponded to the *lep*B and *era* genes, which means that the *rnc* gene was flanked by the *lep*B and *era* genes at the upstream and downstream regions, respectively (**Figure 3**) [46]. It seems that the three genes are organized in one operon, since there was no promoter detected in the upstream of every orf of *lep*B, *rnc,* and *era* genes. The gene organization was similar to that of *Rhodobacter capsulatus* [41]. Based on the information of *rnc* gene organization in *Shewanella* sp. SIB1 genome, the full length of orf of *rnc* gene could be isolated and then used for the expression of recombinant psychrophilic RNase III. The length of the orf of the *rnc* gene was determined to be 678 bp, which produced the recombinant RNase III with a molecular weight of ±24.8 kDa.

Figure 3.
Molecular organization of rnc gene in Shewanella sp. SIB1 genome. The rnc gene is flanked by lepB and era genes at the upstream and downstream regions. It seems that lepB-rnc-era is organized in one operon since there was no promoter sequence was found at the upstream of each gene. Moreover, the rnc-era sequence overlaps with each other (hatched area), while lepB-era (white area) is separated only by one base. Arrows indicate the expression direction [46].

3.2 Expression of recombinant psychrophilic RNase III

To overexpress the recombinant psychrophilic RNase III from *Shewanella* sp. SIB1, the pET28a expression vector, and *Escherichia coli* BL21(DE3) as a host were used in this work. Insertion of the orf of *rnc* gene into the multiple cloning sites of pET28a produces the recombinant protein that is fused with the hexahistidine tag. The resultant plasmid, pET-*rnc,* was then used to transform *Escherichia coli* BL21(DE3). Expression of the recombinant psychrophilic RNase III was induced by isopropyl thio-b-D-galactopyranoside (IPTG).

The result showed that the recombinant psychrophilic RNase III was accumulated in inclusion body form, although the overproduction was shifted at 20°C (**Figure 4**). Several attempts have been implemented to improve the solubility of recombinant

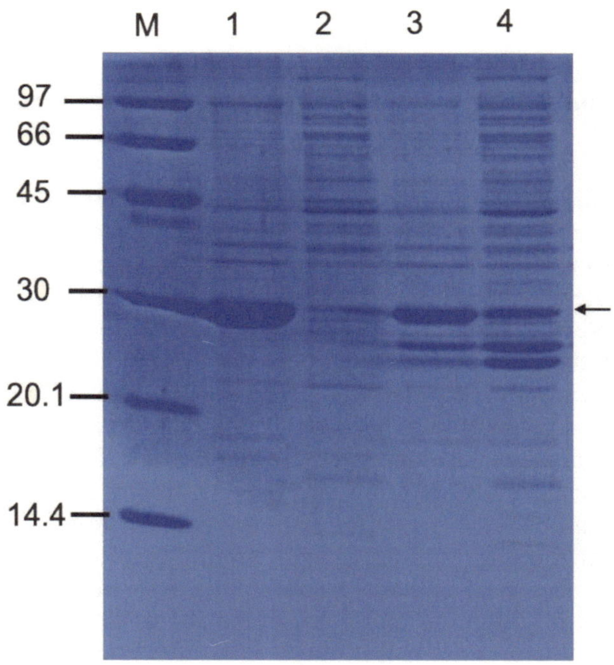

Figure 4.
SDS-PAGE of recombinant psychrophilic RNase III overproduced in Escherichia coli BL21(DE3). Samples were subjected to 15% SDS-PAGE and stained with Coomassie brilliant blue (CBB). Low molecular weight kit (GE Healthcare) (lane M); cell pellet of cell harboring pET-rnc without co-expression with FKBP22 (lane 1); soluble part of cell harboring pET-rnc without co-expression with FKBP (lane 2); cell pellet of cell harboring pET-rnc and FKBP22 (lane 3); and soluble part of cell harboring pET-rnc and FKBP22 (lane 4). Recombinant psychrophilic RNase III was indicated by arrow [46].

psychrophilic RNase III in the *E. coli* system. Shifting of the expression temperatures to 15 and 10°C and adjustment of pH of growing media also did not significantly improve the solubility of recombinant proteins (data not shown).

Another strategy that has been carried out to improve the recombinant psychrophilic RNase III was by co-expression with the chaperone or chaperone-like proteins. Chaperon is a protein that functions for assisting another protein folding. Two types of assisting folding proteins used were GroEL-ES from *Escherichia coli* and FKBP22 from *Shewanella* sp. SIB1 [47]. Among them, co-expression with FKBPP22 successfully improved the solubility of recombinant psychrophilic RNase III (**Figure 2**). FKBP22 belongs to the group of peptidyl-prolyl isomerase (PPIase) that functions for switching *cis*- to *trans*-configuration of proline during polypeptide biosynthesis [47]. This result indicated that strong induction to produce recombinant psychrophilic RNase III might cause the misfolding of the protein. Therefore, during co-expression with FKBP22, it helps to assist the proper folding of the psychrophilic RNase III. Although co-expression with FKBP22 only partly solubilizes the recombinant psychrophilic RNase III, it is sufficient for the biochemical characterization of the recombinant proteins.

Psychrophilic enzymes have unique properties in their folding and activity. Expression of such recombinant psychrophilic enzymes in mesophilic host generally produces misfolding recombinant protein represented by the inclusion bodies formation. Overexpression of recombinant psychrophilic RNase III in *Escherichia coli* has been improved when it was expressed with chaperone-like protein, FKBP22. It is apparently that FKBP22 assists the proper folding of recombinant psychrophilic RNase III.

4. Eukaryote model organism and animal gene transformation Yeast genetics

The yeast *Saccharomyces cerevisiae* is an essential option for expanding breakthrough research in gene cloning in *E. coli*, including eukaryotes. It can be manipulated and cultured using standard techniques applied to unicellular microorganisms. Yeast is a eukaryotes cell whose genetic material is packed into the chromosomes of the membrane-enclosed cell nucleus. In addition, extensive knowledge has been accumulated over the years that yeast has been used as a model system for genetic and biochemical studies. A comprehensive map showing the 17 chromosomes and more than 400 genes is available. The discovery first drove research in this area that yeast genes can be reliably expressed in *E. coli*. Yeast DNA fragments, when cloned into *E. coli* can restore histidine-independent growth of the mutant strain. In another case, a fragment of the yeast chromosome carries the gene for the enzyme that corresponds to the defect in the bacterial strain. Therefore, the yeast HIS3 gene can be expressed in bacterial cells and produce the yeast gene. Usually, wild-type alleles are specified in uppercase, and mutant ones are set in lowercase. Therefore, HIS3 is a wild-type allele, and his3 is a mutant allele that causes histidine dependence. Other yeast genes isolated and used as markers include TRP1, LEU2, URA3, and ARG4. In general, eukaryotic genes have more complex functions than bacterial genes due to introns. Due to the lack of introns, yeast genes may develop easier than other animal cells. An important marker of wild-type yeast attempts to insert exogenous DNA into yeast cells.

4.1 Yeast transformation

Yeast cells are protected by a thick cell wall, a potential barrier to DNA invasion. Removing the cell wall to create protoplasts or spheroplasts increases the chances of genetic transformation. Reseachers adopted this method was adopted and widespread used by these researchers, but some changes have since been have been reported to improve efficiency. This method is based on the technique described initially for protoplast fusion yeast. Yeast cells are recovered in the late stage of growth, the cell wall is weakened with a reducing agent such as mercaptoethanol, and the wall is removed by incubation with an enzyme such as glucanase. Various formulations, such as glucanase enzyme and actinomycete extract have been successfully used. Spheroplasts were then carefully washed with an osmotically equivalent solution of the free buffer and suspended in a solution-containing polyethylene glycol (PEG) and $CaCl_2$ [48]. DNA was added at this stage. For cells to divide, the walls need to be rebuilt. This case requires the cells to be placed in osmotically stabilized agar.

4.2 Gene recognition and gene number regulation

Both plasmid vectors and chromosomal integration are widely used to introduce genes and control copy numbers into *S. cerevisiae*. Each has an important role, and the choice depends on the overall goal (overexpression, tight control of gene number, etc.) [49]. The plasmids used in yeast are far more limited than the *E. coli*. However, plasmids with little copy number control and isolation stability can be a significant problem even in selective media. Homologous recombination is so efficient in *S. cerevisiae* that integrating genes into the genome provides an alternative and simple mechanism for introducing genes. Chromosomal integration also allows the insertion of several identical or different genes. It is critical for the gene expression of regulated metabolic pathways. There are classes of plasmids that replicate independently in yeast: YIp, YAC, YRp, Yep, and YCp [50–53]. *S. Sacevisiae* has a multi-cloning site (MCS) for inserting expression cassettes. The YRp vector originates from replication such as Autonomously Replicating Sequence (ARS) without partition control. However, this plasmid is extremely unstable and is not widely used in metabolic engineering applications. In contrast, the widely used YCp and YEp vectors have been demonstrated in many applications. The YCp vector (centromere/CEN) has an origin of replication; the centromere sequence is maintained at 12 copies per cell and exhibits high isolation stability in selective media. Strong constitutive promoter expression can significantly affect plasmid stability, reduce average copy counts, and overwhelm intracellular metabolic pathways [54]. In extreme cases, the CEN/ARS vector provides overproduction. Due to the general lack of yeast plasmids, very high copy counts were maintained. On the other hand, defective marker promoters lead to increased copy counts [55]. Hundreds of copies have been reported on selective media, but this high copy count is not essential for survival [54]. Generally, such vectors help with the overexpression of product genes rather than metabolic engineering applications [49]. There are 11 classes of animals' homeobox that share homology and function among yeast and animal (**Table 3**). Today, the use of model organisms to replace animal cells is increasing more rapidly due to animal-free thinking in social development. However, cloning and transformation in mammals remain important [51].

Class	Sub Class	Gene	Function	Organism	Ref
ANTP	EuHox	*Hox1, Hox2, Hox3, Hox4, Hox5, Hox6–8, Hox9–15*	Essential for normal T lymphocyte and activated natural killer cell function	*Isodiametra pulchra* (Xenacoelomorpha)	[55, 56]
	SuperHox	*Evx, Meox, Mnx, En, Gbx, Ro, DLx, Nedx, Hex,*	Essential for the development of the pancreas	Zebrafish	[55]
	ParaHox	*Gsx, Pdx, Cdx*	Regulate LGE patterning but oppositely control the balance between proliferation and differentiation in the neuronal progenitor pool.	*Isodiametra pulchra* (Xenacoelomorpha)	[55, 57]
	NK cluster	*Msx, NK4, NK3, Lbx, Tlx, NK1, NK5, NK6, NK7*	Essential for normal heart morphogenesis	*Isodiametra pulchra* (Xenacoelomorpha)	[55, 58]
		Emx, Hlx, Dbx, Barh1, Barx, Bsx, Bari, Vax, Noto, NK2.1, NK2.2, Msxlx, Abox	Regulate postnatal myogenesis, including muscle maintenance during aging and regeneration of acute and chronic muscle injury.	*Isodiametra pulchra* (Xenacoelomorpha)	[55, 59]
PRD	PAX	*Arx, Alx, Hbn, Rax, Otp, Gsc, Otx, Pitx*	Regulate the development of the animal olfactory system	Mouse	[55, 60]
		Pax 1/9, Pon, Pax2/5/8, Pax3/7, Pax4/6/10, Eyg, Pax-alpha,	Involved in the development of the eye structures of *Drosophila*	*Drosophila melanogaster*	[55, 61]
		Vsx, Dmbx, Drgx, Phox, Prop, Prrx, Repo, Shox, Unox, Hopx	Expressed in the presumptive midbrain at early developmental stages, and the hindbrain at later stages, with exclusion from the MHB	*Ciona, Isodiametra pulchra* (Xenacoelomorpha)	[55, 62]
LIM		*Isl, Lmx, Lhx1/5, Lhx2/9, Lhx3/4, Lhx6/8,*	Encoding gene regulation during the pituitary gland, eye, and pancreas, organs assembly that was presumably not present in the common ancestor of vertebrates.	*Isodiametra pulchra* (Xenacoelomorpha) *C. elegans*	[55]
TALE		*Irx, Mkx, Pbx, Meis, Pknox, Tgif*	Plays a critical role in tendon differentiation by regulating type I collagen production in tendon cells.	*Isodiametra pulchra* (Xenacoelomorpha)	[55, 63]

Class	Sub Class	Gene	Function	Organism	Ref
POU		*Pou1, Pou2, Pou3, Pou4,*	Essential for the active maintenance of the differentiated state of a neuron across phylogeny.	*Isodiametra pulchra* (Xenacoelomorpha)	[55, 64]
SINE		*Six1/2, Six3/6, Six4/5,*	Play an essential role in retinal development and influence that these proteins have on cell proliferation and growth	*Isodiametra pulchra* (Xenacoelomorpha)	[55, 65]
CUT		*Cmp, Cux, Onecut*	Developing kidney with expression restricted to the nephrogenic zone.	*Isodiametra pulchra* (Xenacoelomorpha)	[55, 66]
ZF		*Zfhx, Zeb, Tshz*	Specification of individual anterior neural precursors and promotes the expression of *tph* and *synaptotagminB*, required for the differentiation of serotonergic neurons.	*Isodiametra pulchra* (Xenacoelomorpha), sea urchin	[55, 67]
HNF		*Hmbox, HNF*	The inducible protective mechanism that inhibits LPS-induced ROS production and inflammation in EA.hy926 cells by the subsequent inhibition of redox-sensitive NF-κB and MAPK activation.	*Isodiametra pulchra* (Xenacoelomorpha)	[55, 68]
CERS		*Cers*	Synthesize ceramides containing mainly C20–C26 fatty acids, with little or no synthesis of C16- and C18-ceramides	*Isodiametra pulchra* (Xenacoelomorpha)	[55, 69]
PROSS		*Prox*	Expressed in the developing CNS, lens-secreting cone cells of the eye, and midgut. In the mouse, Prox 1 is expressed in many of the same tissues. Young neurons of the subventricular region of the CNS, developing eye lens, and pancreas. Expression is also detected in the developing liver and heart, as well as transiently in the skeletal muscles	Mouse, *Drosophila melanogaster, Isodiametra pulchra* (Xenacoelomorpha)	[55, 70]

Table 3.
Homeobox gene in animal.

4.3 Cloning in animal cells

The development of a vector system for gene transformation in animal cells is under consideration [71]. These vectors are required in biotechnology to synthesize recombinant proteins from genes that are not correctly expressed when cloned in *E. coli* or yeast. Human cloning techniques are sought after by clinical molecular biologists seeking to develop gene therapy techniques: Diseases are treated by introducing the cloned genes into patients [71]. The clinical aspect means that the most excellent attention is paid to the mammalian cloning system, but significant advances have also been made in insects. Cloning insects is fascinating because it uses a new type of vector that we have never encountered.

4.4 Cloning in mammals

Currently, gene cloning in mammals is performed for one of three reasons:
(1) To produce recombinant proteins in mammalian cell culture and related farming techniques. Milk. (2) In gene therapy, human cells are manipulated to treat diseases.
(3) Achieve gene knockout, an important technique used to determine the function of unknown genes. These experiments are usually performed on rodents, such as mice. Viruses as a mammalian clone vector have been known to be the key to cloning mammals for many years. The first cloning experiment with mammalian cells was performed in 1970 using a vector-based on Simian virus 40 (SV40) [72, 73]. The virus can infect several mammalian species following a lysogenic cycle in some hosts and others. SV40 has the same problem as e and has a calicivirus embedded in it. This is because packaging restrictions limit the amount of new DNA inserted into the genome. Therefore, cloning with the SV40 requires replacing one or more of the existing genes with DNA to clone. The original experiment replaced the late gene region segment, but early gene replacement was also an option [73]. However, the discovery of CRISPR/Cas which is based on cloning technology is one of the essential techniques in gene therapy [74].

5. Conclusions

Genes are the universal language that controls the nature of all living things, shared homology among organisms. It is always interesting to reveal the evolution of cloning and gene expression in plant, bacteria, and animal cells. Therefore, with the discovery of genetic engineering, possible to exchange good genetic traits which beneficial for human life. In conclusion, genetic transformation is a genetic engineering technique that can be used to understand the function of a gene or several genes in various events in the life of an organism, both prokaryotes and eukaryotes, so that genetic transformation is carried out for two kinds of purposes, namely scientific purposes to determine the function of certain genes in an organism, and economic goals to improve the quality and productivity of an organism to increase the economic value of an organism. In the future, genetic engineering on prokaryotes and eukaryotes perspective can be used for various purposes in the fields of medicine, agriculture, horticulture, forestry, and food.

Acknowledgements

We thank Badan Penerbit dan Publikasi Universitas Gadjah Mada (BPP UGM) for supporting this publication.

Conflict of interest

The authors declare no conflict of interest.

Appendices and nomenclature

35S	Cauliflower Mozaic Virus 35S promoter
Argonaute	a part of RNA-induced silencing complex (RISC), plays a central role in RNA silencing processes
BIP116b	brassinosteroid Interacting protein 116b
Dicer	human RNase III
Drosha	a class III of RNase III
dsRBD	double-strand binding domain
dsRNA	double-strand RNA
era	era protein-encoding gene
FKBP22	peptidyl-prolyl isomerase protein, a chaperone-like protein from psychrophilic bacterium Shewanella sp. SIB1
GFP	*green fluorescent protein*
GroEL-ES	chaperonin
Hd3a	*heading date 3a*
IPTG	isopropyl thio-b-D-galactopyranoside
lepB	signal peptidase encoding gene
OsKANADI	*Oryza sativa KANADI*
P19	RNA silencing suppressor p19
pACYC-Duet	bacterial expression vector
pET	bacterial expression vector
pRSET	bacterial expression vector
RNase III	ribonuclease III
Rnc	ribonuclease III encoding gene
SDS-PAGE	sodium dodecyl sulfate-polyacrylamide gel electrophoresis
Sh-rnc	riboniclease III encoding gene from Shewanella sp. SIB1
T-DNA	transfer DNA
ANTP	*antennapedia*
ARG4	*argininosuccinate lyase*
Arx	*aristaless related homeobox*
Alx	*aristaless-like homeobox*
Hbn	*homeobrain*
Rax	*retina and anterior neural fold homeobox*
Otp	*orthopedia homeobox*
Gsc	*goosecoid homeobox*
Otx	*orthodenticle homolog*
Pitx	*paired-like homeodomain*
$CaCl_2$	*calcium chloride*
CERS	*ceramide synthase*
Cmp	*collagen-mimetic peptide*
Cux	*cut-like homeobox*
Onecut	*one cut homeobox*
CUT	*cut homeobox*
Emx	*empty spiracles homeobox*

Hlx	*H2.0-like homeobox*
Dbx	*developing brain homeobox*
Barh1	*BarH-like 1 homeobox protein*

Author details

Endang Semiarti[1*], Yekti Asih Purwestri[1,2], Saifur Rohman[3]
and Wahyu Aristyaning Putri[1]

1 Biochemistry Laboratory, Department of Tropical Biology, Universitas Gadjah Mada. Jl. Teknika Selatan, Yogyakarta, Indonesia

2 Research Center for Biotechnology, Universitas Gadjah Mada, Yogyakarta, Indonesia

3 Laboratory of Agricultural Microbiology, Faculty of Agriculture, Department of Agricultural Microbiology, Universitas Gadjah Mada, Yogyakarta, Indonesia

*Address all correspondence to: endsemi@ugm.ac.id

References

[1] Thesleff I. Homeobox genes and growth factors in regulation of craniofacial and tooth morphogenesis. Acta Odontologica Scandinavica. 1995;**53**(3):129-134

[2] Takada S, Hibara K, Ishida T, Tasaka M. The *CUP-SHAPED COTYLEDON1* gene of *Arabidopsis* regulates shoot apical meristem formation. Development. 2001;**128**:1127-1135

[3] Matsuoka M, Ichikawa H, Saito A, Tada Y, Fujimura T, Kano-Murakami Y. Expression of a rice Homeobox gene causes altered morphology of transgenic plants. American Society of Plant Physiologists. 1993;**5**:1039-1048

[4] Kusaba S, Kano-Murakami Y, Matsuoka M, Tamaoki M, Sakamoto T, Yamaguchi I, et al. Alteration of hormone levels in transgenic tobacco plants overexpressing the rice Homeobox gene *OSH1* [Internet]. Plant Physiology. 1998;**116**:471-476. Available from: www.plantphysiol.org

[5] Sato Y, Sentoku N, Miura Y, Hirochika H, Kitano H, Matsuoka M. Loss-of-function mutations in the rice homeobox gene *OSH15* affect the architecture of internodes resulting in dwarf plants. The EMBO Journal. 1999;**18**:992-1002

[6] Lincoln CL, Ong J, Judy Y, Serikawa K, Hakeaibi S. A knottedl-like Homeobox Gene in *Arabidopsis* is expressed in the vegetative meristem and dramatically alters L6af morphology when over-expressed in transgenic plants. The Plant Cell. 1994;**6**:581-590

[7] Byrne ME, Groover AT, Fontana JR, Martienssen RA. Phyllotactic pattern and stem cell fate are determined by the Arabidopsis homeobox gene *BELLRINGER*. Development. 2003;**130**: 3941-3950

[8] Kubo H, AJM P, MGM A, Pereira A, Koornneef M. *ANTHOCYANINLESS2*, a Homeobox gene affecting anthocyanin distribution and root development in Arabidopsis [Internet]. The Plant Cell. 1999;**11**:1217-1226. Available from: www.plantcell.org

[9] Ohgishi M, Oka A, Morelli G, Ruberti I, Aoyama T. Negative autoregulation of the Arabidopsis homeobox gene *ATHB-2*. The Plant Journal. 2021;**25**(4):389-398

[10] Dong Y-H, Yao J-L, Atkinson RG, Putterill JJ, Morris BA, Gardner RC. *MDH1*: An apple homeobox gene belonging to the *BEL1* family. Plant Molecular Biology. 2000;**42**:623-633

[11] Ejaz M, Bencivenga S, Tavares R, Bush M, Sablowski R. *Arabidopsis thaliana* Homeobox Gene *1* controls plant architecture by locally restricting environmental responses. PNAS. 2021;**18**:1-8

[12] Shen B, Sinkevicius KW, Selinger DA, Tarczynski MC. The homeobox gene *GLABRA2* affects seed oil content in Arabidopsis. Plant Molecular Biology. 2006;**60**(3):377-387

[13] Hendelman A, Zebell S, Rodriguez-Leal D, Dukler N, Robitaille G, Wu X, et al. Conserved pleiotropy of an ancient plant homeobox gene uncovered by cis-regulatory dissection. Cell. 2021;**184**(7):1724-1739

[14] Hirakawa Y, Kondo Y, Fukuda H. *TDIF* peptide signaling regulates vascular stem cell proliferation via the *WOX4* homeobox gene in Arabidopsis. Plant Cell. 2010;**22**(8):2618-2629

[15] Janssen B-J, Lund L, Sinha N. Overexpression of a Homeobox Gene,

LeT6, Reveals Indeterminate Features in the *Tomato Compound Leaf 1* [Internet]. 1998;**117**(3):771-786. Available from: www.plantphysiol.org

[16] Lee YH, Oh HS, Cheon CI, Hwang IT, Kim YJ, Chun JY. Structure and expression of the *A. thaliana* homeobox gene *Athb-12*. Biochemical and Biophysical Research Communications. 2001;**284**(1):133-141

[17] Belles-Boix E, Hamant O, Witiak SM, Morin H, Traas J, Pautot V. *KNAT6:* An Arabidopsis homeobox gene involved in meristem activity and organ separation. Plant Cell. 2006;**18**(8):1900-1907

[18] Wang Y, Henriksson E, Söderman E, Henriksson KN, Sundberg E, Engström P. The Arabidopsis homeobox gene, *ATHB16*, regulates leaf development and the sensitivity to photoperiod in Arabidopsis. Developmental Biology. 2003;**264**(1):228-239

[19] Dockx J, Quaedvlieg N, Keultjes G, Kock P, Weisbeek P, Smeekens S. The homeobox gene *ATK1* of *A. thaliana* is expressed in the shoot apex of the seedling and in flowers and inflorescence stems of mature plants. Plant Molecular Biology. 1995;**28**:723-737

[20] Ingram GC, Magnard J-L, Vergne P, Dumas C, Rogowsky PM. ZmOCL1, an *HDGL2* family homeobox gene, is expressed in the outer cell layer throughout maize development. Plant Molecular Biology. 1999;**40**:343-354

[21] Horst NA, Katz A, Pereman I, Decker EL, Ohad N, Reski R. A single homeobox gene triggers phase transition, embryogenesis and asexual reproduction. Nature Plants. 2016;**2**:1-6

[22] Carabelli M, Morellit G, Whitelamt G, Ruberti I, La R, Le SP, et al. Twilight-zone and canopy shade induction of the *Athb-2* homeobox gene in green plants. Developmental Biology Communicated by Walter Journal. 1996;**93**:3530-3535

[23] Bowman J, Eshed Y. Embryonic origin of the shoot apical meristem. Trends in Plant Science. 2000;**5**(3):110-115

[24] Pitzschke A, Persak H. Poinsettia protoplasts-a simple, robust and efficient system for transient gene expression studies [Internet]. Plant Methods. 2012;**8**:2-11. Available from: http://www.plantmethods.com/content/8/1/14

[25] Wang H, Wang W, Zhan J, Huang W, Xu H. An efficient PEG-mediated transient gene expression system in grape protoplasts and its application in subcellular localization studies of flavonoids biosynthesis enzymes. Scientia Horticulturae. 2015;**191**:82-89

[26] Sparkes IA, Runions J, Kearns A, Hawes C. Rapid, transient expression of fluorescent fusion proteins in tobacco plants and generation of stably transformed plants. Nature Protocols. 2006;**1**(4):2019-2025

[27] Xu K, Huang X, Wu M, Wang Y, Chang Y, Liu K, et al. A rapid, highly efficient and economical method of Agrobacterium-mediated in planta transient transformation in living onion epidermis. PLoS ONE. 2014;**9**(1):1-7

[28] Tyurin AA, Suhorukova A, Kabardaeva KV, Goldenkova Pavlova IV. Transient gene expression is an effective experimental tool for the research into the fine mechanisms of plant gene function: Advantages, limitations, and solutions. Plants. 2020;**9**:1-19

[29] Twyman RM, Christou P, Stoger E. Genetic Transformation of Plants and Their Cells. Marcel Dekker Inc.; 2002

[30] Purwestri YA, Ogaki Y, Tsuji H, Shimamoto K. Functional Analysis of *OsKANADI1*, A Florigen *Hd3a* Interacting Protein in Rice (*O. sativa* L.). Indonesian. Journal of Biotechnology. 2012;**17**(2):169-176

[31] Hutanu D. Recent applications of polyethylene glycols (PEGs) and PEG derivatives. Modern Chemistry & Applications. 2014;**02**(02):1-6

[32] Zarrintaj P, Saeb MR, Jafari SH, Mozafari M. Application of compatibilized polymer blends in biomedical fields. In: Compatibilization of Polymer Blends: Micro and Nano Scale Phase Morphologies, Interphase Characterization and Properties. Oxford: Elsevier; 2019. pp. 511-537

[33] Armstrong CL, Petersen WL, Buchholz WG, Bowen BA, Sulc SL. Plant Cell Reports Factors affecting PEG-mediated stable transformation of maize protoplasts. Plant Cell Reports. 1990;**9**:335-339

[34] Liu YC, Vidali L. Efficient polyethylene glycol (PEG) mediated transformation of the moss *Physcomitrella patens*. Journal of Visualized Experiments. 2011;**50**:1-4

[35] Johnson GCM, Carswell GK, Shillito RD. Direct gene transfer via polyethylene. Journal of Cell Culture Method. Springer Verlag; 1989;**12**:127-133

[36] Zhang Y, Su J, Duan S, Ao Y, Dai J, Liu J, et al. A highly efficient rice green tissue protoplast system for transient gene expression and studying light/chloroplast-related processes [Internet]. Plant Method. 2011;**7**(30):1-14. Available from: http://www.plantmethods.com/content/7/1/30

[37] Nicholson AW. Ribonuclease III mechanisms of double-stranded RNA cleavage. Wiley Interdisciplinary Reviews: RNA. 2014;**5**:31-48

[38] Conrad C, Rauhut R. Ribonuclease III: New sense from nuisance [Internet]. The International Journal of Biochemistry & Cell Biology. 2002;**34**(2):116-129. Available from: www.elsevier.com/locate/ijbcb

[39] Altuvia Y, Bar A, Reiss N, Karavani E, Argaman L, Margalit H. In vivo cleavage rules and target repertoire of RNase III in *Escherichia coli*. Nucleic Acids Research. 2018;**46**(19):10380-10394

[40] Sun W, Nicholson AW. Mechanism of action of *E. coli* ribonuclease III. Stringent chemical requirement for the glutamic acid 117 side chain and Mn2+ rescue of the Glu117Asp mutant. Biochemistry. 2001;**40**(16):5102-5110

[41] Wang Z, Hartman E, Roy K, Chanfreau G, Feigon J. Structure of a yeast RNase III dsRBD complex with a noncanonical RNA substrate provides new insights into binding specificity of dsRBDs. Structure. 2011;**19**(7):999-1010

[42] Wilson RC, Doudna JA. Molecular mechanisms of RNA interference. Annual Review of Biophysics. 2013;**42**(1):217-239

[43] Rosano GL, Ceccarelli EA. Recombinant protein expression in *E. coli*: Advances and challenges. Frontiers in Microbiology. Frontiers Research Foundation. 2014;**5**:1-2

[44] Studier FW. Use of bacteriophage T7 lysozyme to improve an inducible T7 expression system. Journal of Molecular Bioengineering. 1991;**219**:37-44

[45] Kato T, Haruki M, Imanaka T, Morikawa M, Kanaya S. Isolation and characterization of psychrotrophic bacteria from oil-reservoir water and oil sands. Applied Microbiology and Biotechnology. 2001;**55**(6):794-800

[46] Feller G. Molecular adaptations to cold in psychrophilic enzymes.

Cellular and Molecular Life Sciences. 2003;**60**:648-662

[47] Rohman S. Gene cloning and biochemical characterization of mesophilic and psychrophilic RNases III. [Osaka]; 2005 [Unpublished]

[48] Suzuki Y, Takano K, Kanaya S. Stabilities and activities of the N- and C-domains of *FKBP22* from a psychrotrophic bacterium overproduced in *E. coli*. FEBS Journal. 2005;**272**(3):632-642

[49] Seresht AK, Nørgaard P, Palmqvist EA, Andersen AS, Olsson L. Modulating heterologous protein production in yeast: The applicability of truncated auxotrophic markers. Applied Microbiology and Biotechnology. 2013;**97**(9):3939-3948

[50] Brown TA. Gene Cloning & DNA Analysis: An Introduction. 7th ed. Manchester: Wiley Blackwell; 2016. pp. 260-266

[51] Joska TM, Mashruwala A, Boyd JM, Belden WJ. A universal cloning method based on yeast homologous recombination that is simple, efficient, and versatile. Journal of Microbiological Methods. 2014;**100**(1):46-51

[52] van Mullem V, Wery M, de Bolle X, Vandenhaute J. Construction of a set of *Saccharomyces cerevisiae* vectors designed for recombinational cloning. Yeast. 2003;**20**(8):739-746

[53] Goudot C, Etchebest C, Devaux F, Lelandais G. The reconstruction of condition-specific transcriptional modules provides new insights in the evolution of yeast AP-1 proteins. PLoS ONE. 2011;**6**(6):1-12

[54] Benders GA, Noskov VN, Denisova EA, Lartigue C, Gibson DG, Assad-Garcia N, et al. Cloning whole bacterial genomes in yeast. Nucleic Acids Research. 2010;**38**(8):2558-2569

[55] Bossier P, Fernandes L, Rocha D, Rodrigues-Pousada C. Overexpression of *YAP2,* coding for a new YAP protein, and YAP1 in *S. cerevisiae* alleviates growth inhibition caused by 1,10-phenanthroline. Journal of Biological Chemistry. 1993;**268**(31):23640-23645

[56] Brauchle M, Bilican A, Eyer C, Baily X, Martinez P, Ladurner P, et al. *Xenacoelomorpha* survey reveals that all 11 animal homeobox gene classes were present in the first bilaterians. Society for. Molecular Biology and Evolution. 2018;**10**(9):1-36

[57] Lawrence A, Sauvageau G, Humphries K, Largman C. The Role of *HOX* Homeobox genes in normal and leukemic hematopoiesis. STEM Cells. 1996;**14**:281-291

[58] Pei Z, Wang B, Chen G, Nagao M, Nakafuku M, Campbell K. Homeobox genes *Gsx1* and *Gsx2* differentially regulate telencephalic progenitor maturation. Proceedings of the National Academy of Sciences of the United States of America. 2011;**108**(4):1675-1680

[59] Lee YM, Park T, Schulz RA, Kim Y. Twist-mediated activation of the *NK4* Homeobox gene in the visceral mesoderm of Drosophila requires two distinct clusters of E-box regulatory elements*. The Journal of Biological Chemistry [Internet]. 1997;**272**(July 11):17531-17541. Available from: http://www.jbc.org/

[60] Makarenkova HP, Meech R. Barx Homeobox family in muscle development and regeneration. In: International Review of Cell and Molecular Biology. Oxford: Elsevier Inc.; 2012. pp. 117-173

[61] Yoshihara SI, Omichi K, Yanazawa M, Kitamura K, Yoshihara Y. *Arx* homeobox

gene is essential for development of mouse olfactory system. Development. 2005;**132**(4):751-762

[62] Aldaz S, Morata G, Azpiazu N. The *Pax-homeobox* gene eyegone is involved in the subdivision of the thorax of Drosophila. Development. 2003;**130**(18):4473-4482

[63] Takahashi T, Holland PWH. Amphioxus and ascidian Dmbx homeobox genes give clues to the vertebrate origins of midbrain development. Development. 2004;**131**(14):3285-3294

[64] Ito Y, Toriuchi N, Yoshitaka T, Ueno-Kudoh H, Sato T, Yokoyama S, et al. The *Mohawk homeobox* gene is a critical regulator of tendon differentiation. Proceedings of the National Academy of Sciences of the United States of America. 2010;**107**(23):10538-10542

[65] Serrano-Saiz E, Leyva-Díaz E, de La Cruz E, Hobert O. BRN3-type *POU* Homeobox genes maintain the identity of mature postmitotic neurons in nematodes and mice. Current Biology. 2018;**28**(17):2813-2823

[66] Kumar JP. The *sine oculis homeobox (SIX)* family of transcription factors as regulators of development and disease. Cellular and Molecular Life Sciences. 2009;**66**(4):565-583

[67] Ledford AW, Brantley JG, Kemeny G, Foreman TL, Quaggin SE, Igarashi P, et al. Deregulated expression of the homeobox gene *Cux-1* in transgenic mice results in downregulation of p27kip1 expression during nephrogenesis, glomerular abnormalities, and multiorgan hyperplasia. Developmental Biology. 2002;**245**(1):157-171

[68] Yaguchi J, Angerer LM, Inaba K, Yaguchi S. Zinc finger homeobox is required for the differentiation of serotonergic neurons in the sea urchin embryo. Developmental Biology. 2012;**363**(1):74-83

[69] Yuan HX, Feng XE, Liu EL, Ge R, Zhang YL, Xiao BG, et al. 5,2′-dibromo-2,4′,5′-trihydroxydiphenylmethanone attenuates LPS-induced inflammation and ROS production in EA.hy926 cells via *HMBOX1* induction. Journal of Cellular and Molecular Medicine. 2019;**23**(1):453-463

[70] Markitantova Y, Simirskii V. Inherited eye diseases with retinal manifestations through the eyes of homeobox genes. International Journal of Molecular Sciences. MDPI AG. 2020;**21**:1-51

[71] Oliver G, Sosa-Pineda B, Geisendorf S, Spana EP, Doe CQ, Gruss P. *Prox 1, a prospero-related homeobox* gene expressed during mouse development. Mechanisms of Development. 1993;**44**(1):3-16

[72] Ayala FJ. Cloning humans? Biological, ethical, and social considerations. Proceedings of the National Academy of Sciences of the United States of America. 2015;**112**(29):8879-8886

[73] Herzberg M, Winocour E. Simian virus 40 deoxyribonucleic acid transcription *In Vitro*: Binding and transcription patterns with a *Mammalian Ribonucleic Acid Polymerase1*. Journal of Virology. 1970;6(5):667-676

[74] Mengstie MA, Wondimu BZ. Mechanism and applications of crispr/cas-9-mediated genome editing. In: Biologics: Targets and Therapy. Vol. 15. Dove Medical Press Ltd.; 2021. pp. 353-361

Section 2

Viral Vectors

Chapter 3

Viral Vectors in Gene Therapy and Clinical Applications

Alexandra L.G. Mahoney, Najah T. Nassif, Bronwyn A. O'Brien and Ann M. Simpson

Abstract

Developments in gene therapy, coupled with advances in genome sequencing and a greater understanding of DNA sequences, have given rise to an exciting area of research. The use of viral vectors in gene therapy has become a very promising and fast-emerging technology over the past few decades. Despite previous setbacks, the approval of viral vector therapies worldwide, with many in late-stage clinical trials has led to a significant increase in research in this area of gene therapy. Retroviral, adenoviral, adeno-associated viral, and lentiviral vectors are all key vectors currently being researched and used in clinical trials. There are many challenges with the use of viral vectors that are yet to be overcome including cost of production, the immune response, and the ability to precisely regulate the expression of the transgene. However, with increased numbers of clinical trials showing efficacy, safety, and growing financial investment, the future use of viral vectors in gene therapy is increasingly promising.

Keywords: gene therapy, viral vector, clinical trials, approved therapies, vector production

1. Introduction

Gene therapy, defined as the delivery of specific genes to a target cell to treat a disorder, is a promising molecular technology that has quickly become a prominent area of research. Clinical disorders that could be treated using gene therapy include severe combined immunodeficiency (SCID), haemophilia, retinitis pigmentosa, diabetes, and various types of cancers [1–3]. With our increasing understanding of gene function and interactions, as well as the greater availability of genome sequencing, our knowledge of how DNA sequences can be used to treat or cure diseases caused by genetic dysfunction has developed greatly.

The delivery of specific genetic material into a host cell requires the use of a vector, or vehicle, for the transfer of a transgene to a specific cell type, by either viral or non-viral means. Techniques for the delivery of non-viral vectors include electroporation, lipofection, and microRNA, which are all useful gene therapy methods as they carry decreased biological risk, offer reduced immunogenicity, and cost less in both money and time to produce when compared to viral vectors [4]. However, the ability of a non-viral vector to enter a cell by transfection is not as efficient as viral vectors, accordingly, research over past decades has been more focused on the use of viral vectors and this is the focus of this review [5].

Common viruses that have been used as vectors include adenovirus, adeno-associated virus, retrovirus, and lentivirus [6–9]. While there have been limitations associated with the use of these viruses, further research, and enhancements in their construction will likely permit their use in a clinical setting. In fact, there are currently several clinical trials using viral vectors in gene therapy for various conditions worldwide [10]. The successful use of viral vectors in late-stage clinical trials and laboratory settings has facilitated growing investment from venture-capital firms and increasing acquisitions of gene therapy start-ups from pharmaceutical companies [11]. The increasing focus on, and investment in viral vectors in gene therapy is a very promising sign for their future use.

This chapter provides a summary of different viral vectors currently being investigated for use in gene therapy. It also provides a review of the different clinical applications of these viral vectors and addresses the advantages and limitations of their use. Successes observed using these vectors and the limitations that this area is currently facing are also discussed.

2. Viral vectors

Viruses have evolved structural characteristics that allow them to efficiently enter a host cell and replicate effectively [12]. We are positioned to exploit these features to produce safe vectors for clinical use, while still maintaining the ability of a virus, carrying a transgene, to enter a host cell. This offers tremendous potential for very impactful therapies for a range of diseases. A viral vector is broadly made-up of three different components, which will vary depending on the type of virus from which it is derived [13]. These essential components include an envelope, the desired transgene (which is encapsulated by the envelope), and a regulatory cassette consisting of a group of genes that control the expression of the transgene. The incorporation of all of these components to form a vector system is outlined in **Figure 1**.

Viral vectors have been used in clinical trials over the past four decades with various levels of success. In 1999, a clinical trial participant died after receiving an adenoviral vector to treat partial ornithine transcarbamylase (OTC) deficiency. The patient suffered a systemic pro-inflammatory response, causing multiple organ system failures [14]. In another clinical trial, success was observed when a patient with X-linked severe combined immunodeficiency (SCID X1) was treated by retrovirus-mediated gene transfer to CD34 bone marrow cells [15]. However, in other patients in the trial, this treatment triggered the development of leukaemia [16]. These negative outcomes reduced both funding and confidence in gene therapy, especially adenoviral and retroviral-based vector systems. Despite this, research has continued to better understand the safety and efficacy of viral vectors to make them a viable clinical option. The viral vectors that have been most intensively researched are retroviral, adenoviral, adeno-associated, and lentiviral vectors.

2.1 Retroviral vectors

Retroviruses possess two copies of single-stranded RNA, coding for the viral proteins; group antigens (gag), DNA polymerase (pol), and the viral envelope (env).

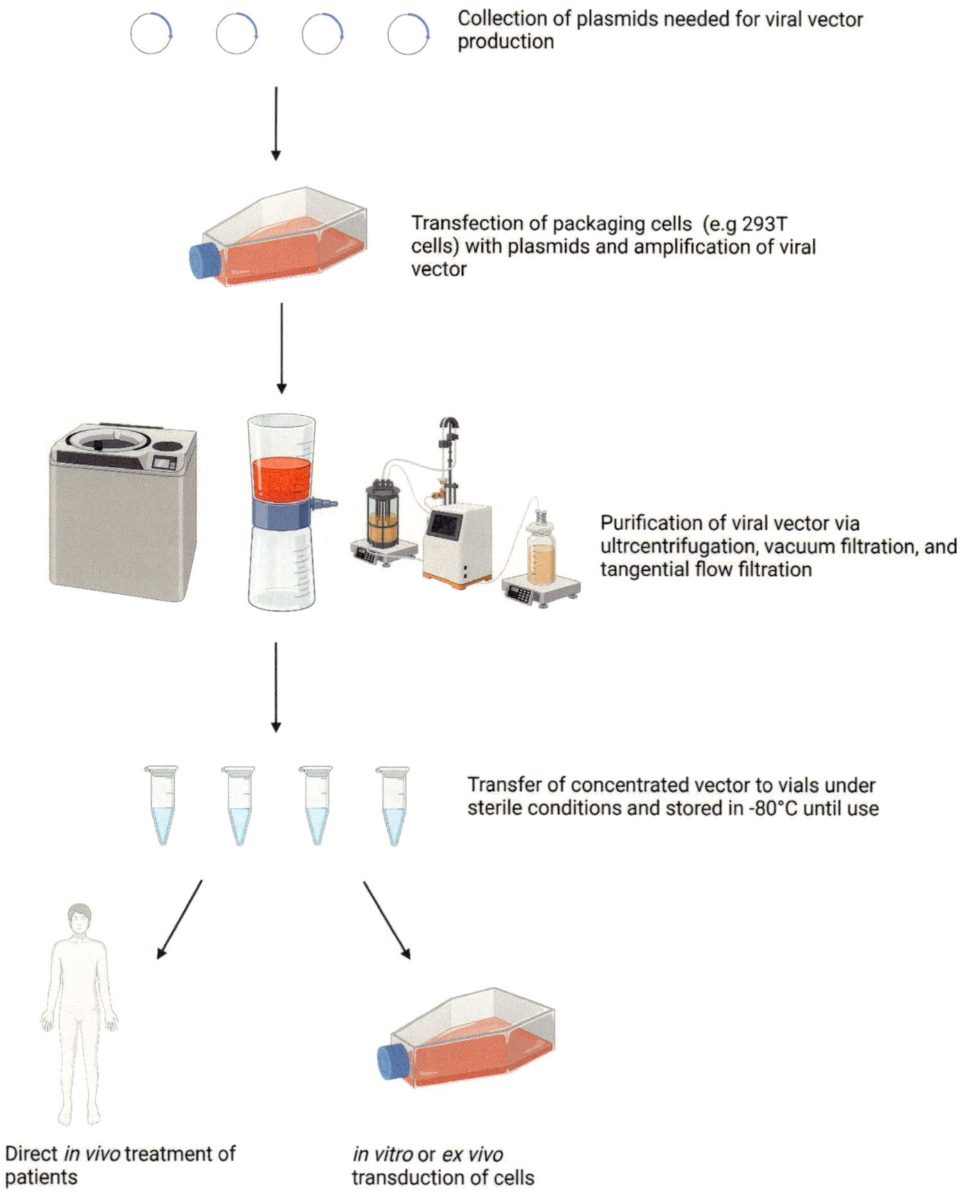

Figure 1.
Production of viral vectors for both in-vivo and in-vitro applications. Plasmid number and packaging cells may differ depending on the type of viral vector being produced. Image created with BioRender (Biorender.com).

The RNA strands are encapsulated by a glycoprotein envelope which allows this virus to enter a target cell. Once internalised, the viral genome integrates within the host DNA, forming a provirus [8]. Viral proteins are then able to be transcribed and translated, after which they exit the cell. Due to their ability to effectively enter a target cell, retroviral vectors are one of the most widely used viral vectors in gene therapy. Retroviral vectors are developed from a disabled murine virus and can only transduce dividing cells [17]. Retroviral vectors have been beneficial in gene therapy as they can integrate into the host cell genome, allowing for sustained gene expression. However, the production of viral proteins poses the risk of insertional mutagenesis occurring, potentially leading to tumour development. This was evident in 2003 when this type of vector was used in a clinical trial for the treatment of (SCID)-X1 disease in which four participants developed leukaemia 3 years after treatment [16]. This was due to the activation of a cellular oncogene during retroviral-vector integration. This raised concerns surrounding the biosafety of this vector and caused a re-evaluation of the use of retroviral vectors in gene therapy, thereby shifting the focus to alternative viral vector systems.

2.2 Adenoviral vectors

Adenoviruses are non-enveloped, double-stranded DNA viruses which are members of the Adenoviridae family [18]. There are at least 47 human adenovirus types, which commonly cause conjunctival and respiratory diseases [6]. Human adenoviruses are ubiquitous in the environment; therefore most people will have immunity to the virus. Infection is usually only mild, but in immunosuppressed individuals, it can be severe. Unlike retroviral vectors, adenoviral vectors can transfer genes to both dividing and non-dividing cells and possess a relatively large cassette capacity (8 kB). They can also be produced in high titres and deliver genes at a high multiplicity of infection [17, 19]. Due to these properties, they have been one of the most common viral vectors used in *in-vivo* experiments and for gene therapy clinical trials. However, adenoviral vectors can elicit a strong inflammatory response due to past exposures generating immunological memory, which can significantly limit their clinical applicability [20]. Additionally, adenoviral vectors cannot integrate into the chromosome of the host, which means the expression of the transgene is episomal and therefore transient. Because of this limitation, adenoviral vectors are not commonly used for disorders that require sustained gene expression but are more frequently used to produce short-term gene expression. For example, adenoviral vectors have applications in cancer research to deliver a suicide gene to kill tumour cells [21].

2.3 Adeno-associated viral vectors

Adeno-associated viruses (AAV) are small, non-enveloped virions containing single-stranded DNA molecules. These viruses are members of the *Dependovirus* genus because they require co-infection with other viruses, and can transduce both dividing and non-dividing cells with long-term expression [22]. Adeno-associated viruses express the viral genes rep (replication), cap (capsid), and aap (assembly) viral genes, but these are removed when developing the AAV vectors, thereby, improving their safety profile [23]. The ability of AAV to enter a host cell and generate recombinant AAV molecules without the aid of viral proteins is a key component favouring their use and distinguishes them from other vector systems. The limited risk of the virus to cause disease and/or adverse events is the main reason why AAV has become an increasingly popular choice over recent decades. The site-specific nature of their integration further increases their safety profile as it limits potential oncogenic consequences. However, these vectors have a limited gene cargo capacity (4.8 kB), and many people have pre-existing antibodies against the variants of AAV, which may have an impact on gene transfer and expression levels [7]. Some serotypes of AAV are unable to reach expression levels high enough to be effective therapeutically, and this is a limitation that needs to be overcome for AAV to be utilised widely for clinical applications.

2.4 Lentiviral vectors

Lentiviruses are RNA viruses that are members of the Retroviridae family. Infection with lentiviruses can lead to many types of diseases, including neurological disorders, arthritis, and immunodeficiency. Lentiviruses have glycoproteins on their surface allowing them to gain entry into a variety of cell types [24]. Like retroviral vectors, they possess the viral genes gag, pol, and env, which allow survival and replication of the lentivirus, as well as the tat and rev genes, which enhance gene transcription and spread of the virus [25]. Being quite a virulent pathogen, fears of a replication-competent vector forming through the use of lentiviral vectors has reduced their applications in the past.

Lentiviral vectors can transduce both dividing and non-dividing cells, thereby making them an ideal choice for a range of gene delivery applications. Additionally, the lentiviral vectors do not elicit a strong immune response, therefore, these are a favourable option for clinical application. These vectors allow for long-term transgene expression as they integrate into the host genome, and insertion is less likely to occur in close proximity to proto-oncogenes, therefore, limiting the risk of insertional mutagenesis [26]. Most lentiviral vectors have been developed from the human immunodeficiency virus (HIV), which has led to some biosafety concerns.

To improve the safety profile of lentiviral vectors, the second-generation vectors have one packaging plasmid which encodes the gag, pol, rev, and tat genes, and the additional accessory virulence factors have been removed. Although the deletion of accessory factors represents a significant improvement to the original vector system, there is still a risk for the generation of a recombinant virus. To combat this, in the third-generation lentiviral vectors the packaging plasmid has been split further, with the gag and pol genes contained in one packaging plasmid, rev in another, and env in a third plasmid [27, 28]. By doing this, the chances of a recombinant virus forming are extremely low. The third-generation vectors are also self-inactivating due to deletions in the 3'LTR in the vector plasmid, thereby, preventing continuous virus replication. The use of a third-generation, self-inactivating lentiviral vector, as opposed to the second-generation vectors, significantly reduces the biosafety risk of viral replication and development of HIV through the removal of the long terminal repeat promoter [9].

3. Applications and clinical use of viral vectors

Over the past four decades, the number of clinical trials using viral vectors for gene therapy has grown significantly. Throughout this time, there have been many significant discoveries, as well as many setbacks. Despite these early obstacles, intensive research in this area has continued, and these efforts have led to the approval of many viral vector-based therapies, with many others currently undergoing late-stage clinical trials [10]. These therapies are predominately focused on treating different cancers, as well as a smaller number focused on the treatment of monogenic, cardiovascular, and infectious diseases. Over the past two decades, over 20 viral vector-based therapies have been approved, 7 of which are adenoviral, adeno-associated, and lentiviral vector-based therapies [29].

3.1 Approved viral-vector therapies

In the early 1990s, an adenoviral vector was approved for use in clinical trials, representing one of the first viral vectors to achieve such approval [30]. Since then, some adenoviral vectors have been approved for widespread use. 'Gendicine' was the first approved viral vector technology, and was approved in 2003 by the China Food and Drug Administration (FDA) to treat patients with head and neck squamous cell carcinoma [31]. Gendicine is a recombinant adenovirus that expresses the tumour-suppressing protein, p53. As of 2020, 30,000 patients had been treated with Gendicine with significantly higher patient response rates observed when it was used in conjunction with chemotherapy, radiotherapy, and other conventional treatments. The clinical outcomes incorporating this viral system with traditional treatments were more efficacious than the use of traditional treatments alone [32]. Many cancerous tumours occur as a result of mutations to the p53 gene, therefore many clinical studies are currently in progress and the use of Gendicine is becoming increasingly widespread for the treatment of other types of cancers, including breast, liver, pancreas, and colorectal cancers [32]. Another adenoviral vector-based therapy, called Oncorine, was approved by the Chinese FDA in 2005 [33]. Oncorine is used to treat late-stage refractory nasopharyngeal cancer and has been very successful when used in conjunction with chemotherapy and radiotherapy. Due to a deletion in the E2B 55K regions, the vector can only infect and replicate in p53 deficient cells, leading to oncolysis of these cells [34].

The AAV vectors have not been intensively researched for as long as the adenoviral vectors, however, they have been extremely successful since their discovery in the 1960s [35]. There have been three AAV vector-based treatments approved, with two of them remaining on the market. Glybera is an AAV vector-based therapy, which was approved by the European Medical Agency in 2012. Glybera delivers lipoprotein lipase to patients who have lipoprotein lipase deficiency [36]. Although this treatment was able to effectively treat the disease, it was not economically viable to maintain it on the market because the incidence of this disorder is one in one million, and consequently it was discontinued in 2017 [37]. Luxturna is another AAV vector therapy that was granted approval by the FDA in the United States in 2017 [38]. It is prescribed for patients with an inherited retinal disease called Lebers congenital amaurosis, which causes progressive blindness. Luxturna is also a very expensive treatment ($425,000 per eye). However, because more people are affected by Lebers congenital amaurosis, the product has remained on the market [39]. Another AAV vector treatment that has been successful, despite being very expensive, is Zolgensma, which is used to treat patients with spinal muscular atrophy. The therapy works by delivering a motor neuron survival transgene to replace the non-functional gene in patients. It was approved in 2019 by the FDA and has seen patients improve to a point where they can walk unsupported, which had not been possible before the advent of this treatment [40].

Similar to AAV vectors, lentiviral vectors have not been researched for as long as other vector systems, but from the time of their first use in clinical trials in 2003 they have been very successful [41]. Kymriah was approved by the FDA in 2017 for the treatment of paediatric relapsed B-cell acute lymphoblastic leukaemia [42]. Kymriah was the first lentiviral vector-based gene therapy treatment and the first chimeric antigen receptor (CAR) T cell immunotherapy. This type of cancer therapy allows the genetic engineering of a patient's own T cells *ex-vivo* to enable them to recognise and eliminate CD19-positive cells. This has been an extremely successful treatment, with patients with lymphoblastic leukaemia achieving remission for a significant amount of time after treatment [43]. Yescarta is another lentiviral vector technology that uses CAR T cell immunotherapy to treat adults with relapsed B cell lymphoma [44]. Yescarta was approved by the FDA in 2017 and has been very effective in treating this disorder.

3.2 Viral vector therapies in clinical trials

There have been over 3000 approved, ongoing, or completed clinical trials involving the use of viral vectors for gene therapy in the past four decades [45]. The range of disorders being researched for treatment development has expanded with continued research success in the area of gene therapy. Clinical trials of gene therapy for many different types of cancers are currently in progress, including head and neck, lung, ovarian, breast, prostate, hepatocellular carcinoma, and melanoma. A number of monogenic diseases have also been investigated, including SCID-X1, ADA-SCID, mucopolysaccharidosis, and Fanconi anaemia, as well as infectious diseases such as HIV and most recently, COVID-19 [45]. Retroviral, adenoviral, adeno-associated, and lentiviral vectors make up over half of the 3000 clinical trials and as stated above have translated into approved therapeutic treatments that have become available on the market. Looking at the trends in the current clinical trial data, much can be deduced regarding the direction of the future of viral vector-based gene therapy (**Figure 2**).

During the 1990s, retroviral vectors were the most common viral vector used in clinical trials for several disorders, including different cancer types, monogenic diseases, and HIV (**Figure 2a**). Despite being the most popular choice of viral vector 30 years ago, the use of retroviral vectors has been steadily declining. This phenomenon is most likely attributed to its inability to be used in non-dividing cells and a significant risk of insertional oncogenesis, leading to cancerous cell formation [15]. Despite the completion of 536 clinical trials using a retroviral vector, this has not resulted in any retroviral vector-based gene therapy being currently available on the market [45]. One treatment, called Strimvelis, was on the market but has since been removed. Strimvelis was approved by the European Medicines Agency in 2016 as a treatment for ADA-SCID using a retroviral vector to deliver adenosine to a patient's cells by *ex-vivo* delivery [46]. However, the development of leukaemia in a patient in 2020 has been reported anecdotally by Orchard Therapeutics, which has since ceased treatment until the risk factors become better understood and can be mitigated. An observational clinical study is currently underway in Italy with 50 patients, which will be conducted for a minimum of 15 years [46]. In order for retroviral vectors to gain greater use in the future, much more research regarding the mechanism of insertional mutagenesis and ways to improve the safety profile is required.

Of all the viral vectors, adenoviral vectors have been most commonly used in clinical trials with 573 either approved, in progress or completed (**Figure 2b**). With two therapies currently on the market for cancer treatment and two more in late-stage clinical trials, adenoviral vector research and gene therapy approaches are demonstrating considerable success [45]. With 70% of the clinical trials being for cancer treatments, adenoviral vectors have become the most popular viral vector used in cancer gene therapy worldwide [45]. Adenoviral vectors are a popular choice for cancer treatment because of their high immunogenicity. While not beneficial in other contexts, the induction of a robust pro-inflammatory response is highly advantageous for cancer treatment [47]. However, like retroviral vectors, adenoviral vector use has declined in the past decade [10]. This may be because of their lack of translation to late-stage clinical trials, and an increase in the use of both adeno-associated and lentiviral vectors in gene therapy clinical trials.

Adeno-associated viral vectors have been used in a limited number of clinical trials, as compared to other vector systems, however, this has not limited their clinical success. The last decade has seen two AAV-vector-based therapies enter the market, as well as a

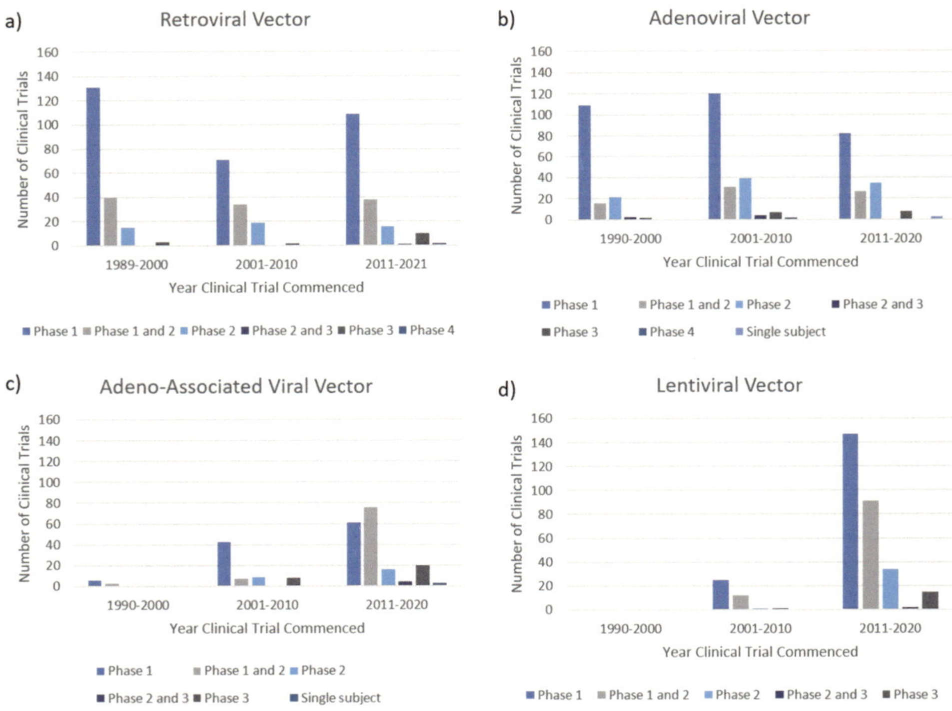

Figure 2.
In-vivo and ex-vivo clinical trials conducted from 1989 to 2021 involving retroviral, adenoviral, adeno-associated, and lentiviral vectors (a–d respectively). Data source from Wiley database on Gene Therapy Trials Worldwide. Available from: http://www.abedia.com/wiley/vectors.php.

sharp increase in phase I and phase II clinical trials (**Figure 2c**). Although many of the AAV vector phase I trials did not begin as early as the other vector trials, they now have the most phase III trials approved, ongoing, or completed as of 2021 (**Figure 2c**). AAV vectors have also shown clinical efficacy in a range of diseases, including antitrypsin deficiency, ocular diseases, and haemophilia [48–51]. Seeing the strides AAV vectors have made in only the past two decades, they appear to be a promising technology for future use. Another promising technology when reviewing clinical trial data is lentiviral vectors. Lentiviral vectors have had the greatest number of clinical trials approved, ongoing, or completed in the past decade despite having the smallest number before 2010 (**Figure 2d**). Some disorders that lentiviral vector use is primarily focused on include cancers, β-thalassemia, HIV, and Fanconi Anaemia. The benefits of using a lentiviral vector over a retroviral vector for transgene delivery is that they can transduce slow dividing or non-dividing cells and seem to have less affinity for integration into oncogenetic sites, especially the self-inactivating, third-generation lentiviral vectors [52, 53]. These lentiviral vectors with a strong promotor largely mitigate the risk of insertional mutagenesis, however, this risk is not eliminated completely. A self-inactivating lentiviral vector has been used in clinical trials for the treatment of HIV with a total of 65 patients treated with the vector and no adverse events reported for more than 8 years after vector infusion [41]. Analysing both the limited numbers of adverse events and the successful clinical trial data over the past three decades reveals that both AAV and lentiviral vectors are favourable gene therapy technologies for the future.

4. Concerns facing viral vector-based gene therapy

Despite their growing success in gene therapy clinical trials, there are still many issues that viral vector technology will need to overcome to be accepted as a widespread therapeutic option. Key areas of concern with the use of viral vectors are the induction of an immune response when delivered *in-vivo*, determining the optimal therapeutic dose required, the cost of production, and the precise regulation of transgene expression levels.

4.1 Immune response

The immunogenicity of a viral vector is measured both quantitatively by the magnitude of the immune response over time, and qualitatively by the types of immune responses that are initiated [54]. Many factors determine the immunogenicity of a vector, and it varies greatly depending on the structure of the viral vector system. It is crucial to understand the interaction of the vector with the immune system before entering clinical trials as the occurrence of a severe immune response upon injection can result in many severe complications, and, in some instances, death [55].

Adenoviral and adeno-associated viral vectors are of particular immunological concern. The prevalence of different adenovirus serotypes varies regionally. For example, serotype 5 (Ad5) has a 50% prevalence in America, but in Africa, this approximates 100% [56]. Despite such variations, adenoviruses are generally prevalent in the environment and are highly immunogenic, which can present concerns when administering vectors using the same serotype [57]. If a patient has already been exposed to the serotype used in the therapeutic vector, this is likely to cause a robust immune response characterised by a rapid influx of pre-existing neutralising antibodies to the injection site, thereby reducing the therapeutic dose and limiting the ability of the vector to exert its clinical effect, and causing safety concerns for the patient due to complement activation and resultant inflammation [55]. This is a similar situation for AAV as approximately 80% of the worldwide population has already been exposed to an AAV serotype [58]. Previous exposure to serotypes will prove to be a major hurdle to overcome in clinical trials for both adenoviral and AAV vectors. In some cases, however, a highly immunogenic vector can be beneficial for the treatment of certain disorders. Adenoviral vectors are the most common vector for cancer therapy mostly due to their highly immunogenic nature. Triggering an anti-tumour response through oncolytic adenovirus treatment has proven to show some success in treating cancerous tumours with two approved cancer treatments on the market [59].

Lentiviral vectors have a very favourable immunogenic profile, as compared to adenoviral vectors, and this is a notable reason why they have been a popular vector choice in the past decade. Lentiviral infection in humans is quite limited, and, therefore, only a small percentage of individuals will carry pre-existing antibodies to the virus. Additionally, in many lentiviral vector systems, the original viral envelope for the Vesicular Stomatitis Virus envelope glycoprotein (VSV-G) has been substituted [24]. Lentiviral vectors trigger long-lasting T-cell immunity, without causing an adverse vector-specific immunity or inflammatory reaction [60], thereby favouring clinical applications.

4.2 Cost of production

The cost to produce a viral vector is an important consideration if the end goal is the clinical application [61]. There are many costs to consider in the production of a viral vector system, including equipment, laboratory material, purification, storage, and the amount of labour needed. As exemplified by the AAV vector-based gene therapy, Glybera, if the product is too expensive to produce and the number of patients affected by the disorder is too low, it may not be economically feasible for the product to stay on the market. A major factor in the cost of production of a vector is the dosage required for one patient. For example, a low inoculation dose can offset a large production cost [62]. One way to lower the cost and time to produce a large amount of the vector is for the vector to have a high titre level. Adenoviral vectors are very efficient at gene transfer, so the titres for these vectors are very high, which is beneficial when produced on a large scale [61]. Adenoviral vectors are a very popular choice for vaccinations, and this aspect of their high titre capability is part of the reason for their popularity. Overall, the cost to produce viral vectors is a significant hurdle that will have to be overcome if they are to be used on a commercial scale. There is currently significant research dedicated to streamlining the process of vector production to lower the cost and time required for production and to allow production in low-resource areas. This discussion is beyond the scope of this review; however, it has been considered elsewhere [63–65].

4.3 Expression of the transgene

Another consideration for viral vector use is the delivery of the transgene and to what degree this process can be controlled. The ability to deliver a specific gene to a cell has proven to be a very effective therapeutic treatment, however, if this does not occur in a regulated manner, it can be detrimental to the patient, especially in cases of random integration. Transgene expression seems at times to be unpredictable, with research showing that in some instances genetic variation can influence expression [66]. Depending on the condition, the transgene will need to be expressed at different levels and potentially only in specific areas or cell types. To control the expression of transgenes and combat unpredictability, strategies such as the use of tissue-specific promoters and self-inactivation have been implemented. Tissue-specific promoters restrict the expression of the transgene to certain cell types only, thus limiting widespread expression. This is ideal when used therapeutically to target a specific cell or tissue type and avoid expression in non-target cells or tissues [67]. Furthermore, as seen in the third-generation lentiviral vectors, a self-inactivating mechanism has been incorporated. Modification in the 3′ long terminal repeat prevents continued expression after one round of integration, effectively allowing the amount of transgene expression to be controlled with the dose of the vector [28]. Despite these positive outcomes, additional research that will enable tightly regulated transgene expression is still required.

5. Conclusions

Viral vector-based gene therapy has made very encouraging strides over the past two decades, suggesting there is a positive future for this therapy in medicine. As reported by IQVIA, the first half of 2021 saw a record amount invested into biopharmaceutical companies by venture capital firms, with cell and gene therapies attracting a significant amount of this investment [11]. The value of pharmaceutical mergers and acquisitions (M&A) in 2021 showed a stark increase from the year before many of which were viral vector and gene therapy-based deals [11, 68]. While these are promising statistics for the future of viral vector use, the concerns facing this method of gene therapy still stand and will require a considerable amount of research to overcome them. Moving forward, considering both the clinical trial data and the drawbacks of each viral vector, it seems lentiviral and adeno-associated viral vectors are the most favourable options to focus research on in the future. With limited adverse reactions and favourable immunogenic profiles, these viral vectors have the potential to be a key treatment in modern medicine.

Acknowledgements

A.L.G. Mahoney was supported by an Australian Government Research Training Stipend.

Conflict of interest

The authors declare no conflict of interest.

Author details

Alexandra L.G. Mahoney, Najah T. Nassif, Bronwyn A. O'Brien and Ann M. Simpson*
School of Life Sciences, University of Technology Sydney, Sydney, Australia

*Address all correspondence to: ann.simpson@uts.edu.au

References

[1] Cavazzana-Calvo M, Hacein-Bey S, de Saint Basile G, Gross F, Yvon E, Nusbaum P, et al. Gene therapy of human severe combined immunodeficiency (SCID)-X1 disease. Science. 2000;**288**: 669-672. DOI: 10.1126/science.288. 5466.669

[2] Acland GM, Aguirre GD, Ray J, Zhang Q, Aleman TS, Cideciyan AV, et al. Gene therapy restores vision in a canine model of childhood blindness. Nature Genetics. 2001;**28**:92-95. DOI: 10.1038/ ng0501-92

[3] Nathwani AC, Reiss UM, Tuddenham EG, Rosales C, Chowdary P, McIntosh J, et al. Long-term safety and efficacy of factor IX gene therapy in hemophilia B. New England Journal of Medicine. 2014;**371**:1994-2004. DOI: 10.1056/NEJMoa1407309

[4] Ramamoorth M, Narvekar A. Non viral vectors in gene therapy-an overview. Journal of Clinical and Diagnostic Research: JCDR. 2015;**9**:GE01-GE06. DOI: 10.7860/JCDR/2015/10443.5394

[5] Wu P, Chen H, Jin R, Weng T, Ho JK, You C, et al. Non-viral gene delivery systems for tissue repair and regeneration. Journal of Translational Medicine. 2018;**16**:1-20. DOI: 10.1186/ s12967-018-1402-1

[6] Wold WSM, Toth K. Adenovirus vectors for gene therapy, vaccination and cancer gene therapy. Current Gene Therapy. 2013;**13**:421-433

[7] Zaiss A, Muruve D. Immunity to adeno-associated virus vectors in animals and humans: A continued challenge. Gene Therapy. 2008;**15**:808-816. DOI: 10.1038/gt.2008.54

[8] Kurian K, Watson C, Wyllie A. Retroviral vectors. Molecular Pathology. 2000;**53**:173. DOI: 10.1136/mp.53.4.173

[9] Buchschacher GL Jr, Wong-Staal F. Development of lentiviral vectors for gene therapy for human diseases. Blood, The Journal of the American Society of Hematology. 2000;**95**:2499-2504. DOI: 10.1182/blood.V95.8.2499

[10] Ginn SL, Amaya AK, Alexander IE, Edelstein M, Abedi MR. Gene therapy clinical trials worldwide to 2017: An update. The Journal of Gene Medicine. 2018;**20**:e3015. DOI: 10.1002/jgm.3015

[11] IQVIA. IQVIA Pharma Deals 2021 [Internet]. 2021. Available from: https://www.iqvia.com/library/white-papers/iqvia-pharma-deals-half-year-review-of-2021 [Accessed: 09 December 2021]

[12] Bamford DH, Grimes JM, Stuart DI. What does structure tell us about virus evolution? Current Opinion in Structural Biology. 2005;**15**:655-663. DOI: 10.1016/j. sbi.2005.10.012

[13] Kay MA, Glorioso JC, Naldini L. Viral vectors for gene therapy: The art of turning infectious agents into vehicles of therapeutics. Nature Medicine. 2001;**7**:33-40. DOI: 10.1038/83324

[14] Raper SE, Chirmule N, Lee FS, Wivel NA, Bagg A, Gao G-p, et al. Fatal systemic inflammatory response syndrome in a ornithine transcarbamylase deficient patient following adenoviral gene transfer. Molecular Genetics and Metabolism. 2003;**80**:148-158. DOI: 10.1016/j. ymgme.2003.08.016

[15] Hacein-Bey-Abina S, Von Kalle C, Schmidt M, McCormack M, Wulffraat N,

Leboulch PA, et al. LMO2-associated clonal T cell proliferation in two patients after gene therapy for SCID-X1. Science. 2003;**302**:415-419. DOI: 10.1126/science.1088547

[16] Kohn D, Mitsuya H, Ballow M, Selegue J, Barankiewicz J, Cohen A, et al. Establishment and characterization of adenosine deaminase-deficient human T cell lines. The Journal of Immunology. 1989;**142**:3971-3977

[17] Morgan RA, Anderson WF. Human gene therapy. Annual Review of Biochemistry. 1993;**62**:191-217. DOI: 10.1146/annurev.bi.62.070193.001203

[18] Randrianarison-Jewtoukoff V, Perricaudet MJB. Recombinant adenoviruses as vaccines. Biologicals. 1995;**23**:145-157. DOI: 10.1006/biol.1995.0025

[19] Volpers C, Kochanek S. Adenoviral vectors for gene transfer and therapy. The Journal of Gene Medicine. 2004;**6**:S164-S171. DOI: 10.1002/jgm.496

[20] Zhou HS, Liu DP, Liang CC. Challenges and strategies: The immune responses in gene therapy. Medical Research Reviews. 2004;**24**:748-761. DOI: 10.1002/med.20009

[21] Tazawa H, Kagawa S, Fujiwara T. Advances in adenovirus-mediated p53 cancer gene therapy. Expert Opinion on Biological Therapy. 2013;**13**:1569-1583. DOI: 10.1517/14712598.2013.845662

[22] Xu R, Li H, Lai-yin T, Hsiang-fu K, Lu H, Lam K. Diabetes gene therapy: Potential and challenges. Current Gene Therapy. 2003;**3**:65-82. DOI: 10.2174/1566523033347444

[23] Worgall S, Crystal RG. Gene therapy. In: Lanza R, Langer R, Vacanti JP, Atala A, editors. Principles of Tissue Engineering. Cambridge, MA, USA: Academic Press; 2020. 34: 657-686. DOI: 10.1016/B978-0-12-398358-9.00034-3

[24] Hastie E, Cataldi M, Marriott I, Grdzelishvili VZ. Understanding and altering cell tropism of vesicular stomatitis virus. Virus Research. 2013;**176**:16-32

[25] O'Connor D. Introduction to gene and stem-cell therapy. In: Molecular and Cellular Therapies for Motor Neuron Diseases. Amsterdam, Netherlands: Elsevier; 2017. pp. 141-165. DOI: 10.1016/j.virusres.2013.06.003

[26] Ronen K, Negre O, Roth S, Colomb C, Malani N, Denaro M, et al. Distribution of lentiviral vector integration sites in mice following therapeutic gene transfer to treat β-thalassemia. Molecular Therapy. 2011;**19**:1273-1286. DOI: 10.1038/mt.2011.20

[27] Vannucci L, Lai M, Chiuppesi F, Ceccherini-Nelli L, Pistello M. Viral vectors: A look back and ahead on gene transfer technology. New Microbiologica. 2013;**36**:1-22

[28] Dull T, Zufferey R, Kelly M, Mandel R, Nguyen M, Trono D, et al. A third-generation lentivirus vector with a conditional packaging system. Journal of Virology. 1998;**72**:8463-8471. DOI: 10.1128/JVI.72.11.8463-8471.1998

[29] Ma C-C, Wang Z-L, Xu T, He Z-Y, Wei Y-Q. The approved gene therapy drugs worldwide: From 1998 to 2019. Biotechnology Advances. 2020;**40**:107502. DOI: 10.1016/j.biotechadv.2019.107502

[30] Zabner J, Couture LA, Gregory RJ, Graham SM, Smith AE, Welsh MJ. Adenovirus-mediated gene transfer

transiently corrects the chloride transport defect in nasal epithelia of patients with cystic fibrosis. Cell. 1993;**75**:207-216. DOI: 10.1016/0092-8674(93)80063-K

[31] Zhang W-W, Li L, Li D, Liu J, Li X, Li W, et al. The first approved gene therapy product for cancer Ad-p53 (Gendicine): 12 years in the clinic. Human Gene Therapy. 2018;**29**:160-179. DOI: 10.1089/hum.2017.218

[32] Xia Y, Li X, Sun W. Applications of recombinant adenovirus-p53 gene therapy for cancers in the clinic in China. Current Gene Therapy. 2020;**20**:127-141. DOI: 10.2174/1566523220999200731003206

[33] Liang M. Oncorine, the world first oncolytic virus medicine and its update in China. Current Cancer Drug Targets. 2018;**18**:171-176. DOI: 10.2174/1568009618666171129221503

[34] Wirth T, Parker N, Ylä-Herttuala S. History of gene therapy. Gene. 2013;**525**:162-169. DOI: 10.1016/j.gene.2013.03.137

[35] Atchison R, Casto BC, Hammon WM. Electron microscopy of adenovirus-associated virus (AAV) in cell cultures. Virology. 1996;**29**:353-357. DOI: 10.1016/0042-6822(66)90045-6

[36] Keeler AM, Flotte TR. Recombinant adeno-associated virus gene therapy in light of Luxturna (and Zolgensma and Glybera): Where are we, and how did we get here? Annual Review of Virology. 2019;**6**:601-621. DOI: 10.1146/annurev-virology-092818-015530

[37] Gaudet D, Méthot J, Déry S, Brisson D, Essiembre C, Tremblay G, et al. Efficacy and long-term safety of alipogene tiparvovec (AAV1-LPL S447X) gene therapy for lipoprotein lipase deficiency: An open-label trial. Gene Therapy. 2013;**20**:361-369. DOI: 10.1038/gt.2012.43

[38] Russell S, Bennett J, Wellman JA, Chung DC, Yu Z-F, Tillman A, et al. Efficacy and safety of voretigene neparvovec (AAV2-hRPE65v2) in patients with RPE65-mediated inherited retinal dystrophy: A randomised, controlled, open-label, phase 3 trial. The Lancet. 2017;**390**:849-860. DOI: 10.1016/S0140-6736(17)31868-8

[39] Bulaklak K, Gersbach CA. The once and future gene therapy. Nature Communications. 2020;**11**:1-4. DOI: 10.1038/s41467-020-19505-2

[40] Broekhoff TF, Sweegers CC, Krijkamp EM, Mantel-Teeuwisse AK, Leufkens HG, Goettsch WG, et al. Early Cost-effectiveness of onasemnogene abeparvovec-xioi (zolgensma) and nusinersen (spinraza) treatment for spinal muscular atrophy i in the Netherlands with relapse scenarios. Value in Health. 2021;**24**:759-769. DOI: 10.1016/j.jval.2020.09.021

[41] McGarrity GJ, Hoyah G, Winemiller A, Andre K, Stein D, Blick G, et al. Patient monitoring and follow-up in lentiviral clinical trials. The Journal of Gene Medicine. 2013;**15**:78-82. DOI: 10.1002/jgm.2691

[42] Bach PB, Giralt SA, Saltz LB. FDA approval of tisagenlecleucel: Promise and complexities of a $475 000 cancer drug. Journal of the American Medical Association. 2017;**318**:1861-1862. DOI: 10.1001/jama.2017.15218

[43] Ali S, Kjeken R, Niederlaender C, Markey G, Saunders TS, Opsata M, et al. The European medicines agency review of Kymriah (Tisagenlecleucel) for the treatment of acute lymphoblastic leukemia and diffuse large B-cell

lymphoma. The Oncologist. 2020;**25**:e321-e327. DOI: 10.1634/theoncologist.2019-0233

[44] Chow VA, Shadman M, Gopal AK. Translating anti-CD19 CAR T-cell therapy into clinical practice for relapsed/refractory diffuse large B-cell lymphoma. Blood Spotlight. 2018;**132**:777-781. DOI: 10.1182/blood-2018-04-839217

[45] The Journal of Gene Medicine. Gene Therapy Clinical Trials Worldwide [Internet]. 2021. Available from: https://a873679.fmphost.com/fmi/webd/GTCT [Accessed: 03 November 2021]

[46] Stirnadel-Farrant H, Kudari M, Garman N, Imrie J, Chopra B, Giannelli S, et al. Gene therapy in rare diseases: The benefits and challenges of developing a patient-centric registry for Strimvelis in ADA-SCID. Orphanet Journal of Rare Diseases. 2018;**13**:1-10. DOI: 10.1186/s13023-018-0791-9

[47] Shaw AR, Suzuki M. Immunology of adenoviral vectors in cancer therapy. Molecular Therapy-Methods & Clinical Development. 2019;**15**:418-429. DOI: 10.1016/j.omtm.2019.11.001

[48] Mueller C, Gernoux G, Gruntman AM, Borel F, Reeves EP, Calcedo R, et al. 5 year expression and neutrophil defect repair after gene therapy in alpha-1 antitrypsin deficiency. Molecular Therapy. 2017;**25**:1387-1394. DOI: 10.1016/j.ymthe.2017.03.029

[49] Bennett J, Wellman J, Marshall KA, McCague S, Ashtari M, DiStefano-Pappas J, et al. Safety and durability of effect of contralateral-eye administration of AAV2 gene therapy in patients with childhood-onset blindness caused by RPE65 mutations: A follow-on phase 1 trial. The Lancet. 2016;**388**:661-672. DOI: 10.1016/S0140-6736(16)30371-3

[50] George LA, Sullivan SK, Giermasz A, Rasko JE, Samelson-Jones BJ, Ducore J, et al. Hemophilia B gene therapy with a high-specific-activity factor IX variant. New England Journal of Medicine. 2017;**377**:2215-2227. DOI: 10.1056/NEJMoa1708538

[51] George LA, Ragni MV, Samelson-Jones BJ, Cuker A, Runoski AR, Cole G, et al. Spk-8011: Preliminary results from a phase 1/2 dose escalation trial of an investigational AAV-mediated gene therapy for hemophilia A. Blood. 2017;**130**:604. DOI: 10.1182/blood.V130.Suppl_1.604.604

[52] Milone MC, O'Doherty U. Clinical use of lentiviral vectors. Leukemia. 2018;**32**:1529-1541. DOI: 10.1038/s41375-018-0106-0

[53] Zufferey R, Dull T, Mandel RJ, Bukovsky A, Quiroz D, Naldini L, et al. Self-inactivating lentivirus vector for safe and efficient in vivo gene delivery. Journal of Virology. 1998;**72**:9873-9880. DOI: 10.1128/JVI.72.12.9873-9880.1998

[54] Bouwman W, Verhaegh W, Holtzer L, van de Stolpe A. Measurement of cellular immune response to viral infection and vaccination. Frontiers in Immunology. 2020;**11**:2719. DOI: 10.3389/fimmu.2020.575074

[55] Huang X, Yang Y. Innate immune recognition of viruses and viral vectors. Human Gene Therapy. 2009;**20**:293-301. DOI: 10.1089/hum.2008.141

[56] Lynch JP III, Kajon AE. Adenovirus: Epidemiology, global spread of novel serotypes, and advances in treatment and prevention. Seminars in Respiratory and Critical Care Medicine, Thieme Medical Publishers. 2016;**37**:586-602. DOI: 10.1055/s-0036-1584923

[57] M.J. Imperiale, Keeping adenovirus away from the liver, Cell Host & Microbe.

2008;3:119-120. DOI: doi.org/10.1016/j.chom.2008.02.007

[58] Calcedo R, Vandenberghe LH, Gao G, Lin J, Wilson JM. Worldwide epidemiology of neutralizing antibodies to adeno-associated viruses. The Journal of Infectious Diseases. 2009;**199**:381-390. DOI: 10.1086/595830

[59] Bartlett DL, Liu Z, Sathaiah M, Ravindranathan R, Guo Z, He Y, et al. Oncolytic viruses as therapeutic cancer vaccines. Molecular Cancer. 2013;**12**:1-16. DOI: 10.1186/1476-4598-12-103

[60] Buffa V, Negri DR, Leone P, Borghi M, Bona R, Michelini Z, et al. Evaluation of a self-inactivating lentiviral vector expressing simian immunodeficiency virus gag for induction of specific immune responses in vitro and in vivo. Viral Immunology. 2006;**19**:690-701. DOI: 10.1089/vim.2006.19.690

[61] Kallel H, Kamen AA. Large-scale adenovirus and poxvirus-vectored vaccine manufacturing to enable clinical trials. Biotechnology Journal. 2015;**10**:741-747. DOI: 10.1002/biot.201400390

[62] Nascimento I, Leite L. Recombinant vaccines and the development of new vaccine strategies. Brazilian Journal of Medical and Biological Research. 2012;**45**:102-1111. DOI: 10.1590/S0100-879X2012007500142

[63] Merten O-W, Schweizer M, Chahal P, Kamen AA. Manufacturing of viral vectors for gene therapy: Part I. Upstream processing. Pharmaceutical Bioprocessing. 2014;**2**:183-203. DOI: 10.4155/pbp.14.16

[64] Chen K-D, Wu X-X, Yu D-S, Ou H-L, Li Y-H, Zhou Y-Q, et al. Process optimization for the rapid production of adenoviral vectors for clinical trials in a disposable bioreactor system. Applied Microbiology and Biotechnology. 2018;**102**:6469-6477. DOI: 10.1007/s00253-018-9091-5

[65] Sharon D, Kamen A. Advancements in the design and scalable production of viral gene transfer vectors. Biotechnology and Bioengineering. 2018;**115**:25-40. DOI: 10.1002/bit.26461

[66] Suwanmanee T, Ferris MT, Hu P, Gui T, Montgomery SA, de Villena FP-M, et al. Toward personalized gene therapy: Characterizing the host genetic control of lentiviral-vector-mediated hepatic gene delivery. Molecular Therapy-Methods & Clinical Development. 2017;**5**:83-92. DOI: 10.1016/j.omtm.2017.03.009

[67] Zheng C, Baum BJ. Evaluation of promoters for use in tissue-specific gene delivery. Gene Therapy Protocols. 2008;**434**:205-219. DOI: 10.1007/978-1-60327-248-3_13

[68] Biopharmaceutical Dealmakers & DealForma, M&As make their mark [Internet]. 2020. Available from: https://www.nature.com/articles/d43747-020-00841-7 [Accessed: 10 Decemeber 2021]

Section 3

CRISPR Technology

Chapter 4

CRISPR Technology: Emerging Tools of Genome Editing and Protein Detection

Rita Lakkakul and Pradip Hirapure

Abstract

CRISPR technology has seen rapid development in applications ranging from genomic and epigenetic changes to protein identification throughout the last decade. The clustered regularly interspaced short palindromic repeats (CRISPR) and CRISPR-associated (Cas) protein systems have transformed the ability to edit, control the genomic nucleic acid and non-nucleic acid target such as detection of proteins. CRISPR/Cas systems are RNA-guided endonucleases exhibiting distinct cleavage activities deployed in the development of analytical techniques. Apart from genome editing technology, CRISPR/Cas has also been incorporated in amplified detection of proteins, transcriptional modulation, cancer biomarkers, and rapid detection of POC (point of care) diagnostics for various diseases such as Covid-19. Current protein detection methods incorporate sophisticated instrumentation and extensive sensing procedures with less reliable, quantitative, and sensitive detection of proteins. The precision and sensitivity brought in by CRISPR-dependent detection of proteins will ensure the elimination of current impediments. CRISPR-based amplification strategies have been used for accurate estimation of proteins including aptamer-based assay, femtomolar detection of proteins in living cells, immunoassays, and isothermal proximal assay for high throughput. The chapter will provide a comprehensive summary of key developments in emerging tools of genome editing and protein detection deploying CRISPR technology, and its future perspectives will be discussed.

Keywords: CRISPR/Cas, genome editing, protein detection, CRISPR technology, anti-CRISPR

1. Introduction

The clustered regularly interspaced short palindromic repeats (CRISPR) and CRISPR-associated (Cas) protein modules are found to be a part of adaptive antivirus defense systems in archaea and bacteria and mediate immunity by a three-stage process called adaption, processing of the primary transcript, and interference. These systems incorporate fragments of foreign DNA (known as spacers) into CRISPR cassettes, then transcribe the CRISPR arrays including the spacers, and process them to make a guide crRNA or the clustered regularly interspaced short palindromic repeats ribonucleic acid (CRISP RNA) which is employed to specifically target and cleave the genome of the cognate virus or plasmid. Earlier classic methods such as zinc finger motif, meganucleases, and transcription activator-like effector nucleases were deployed for genome editing but due to its prerequisite for different fusion proteins, the technology raised hurdles in its applicability. The characteristic feature of single guide RNA of CRISPR to regulate Cas protein to target specific gene sequence is highly advantageous to overcome the barriers posing from classic methods. Proteins cas1 and cas2 genes are found to be the core and active part of the information processing subsystems of the three distinct types of CRISPR/Cas

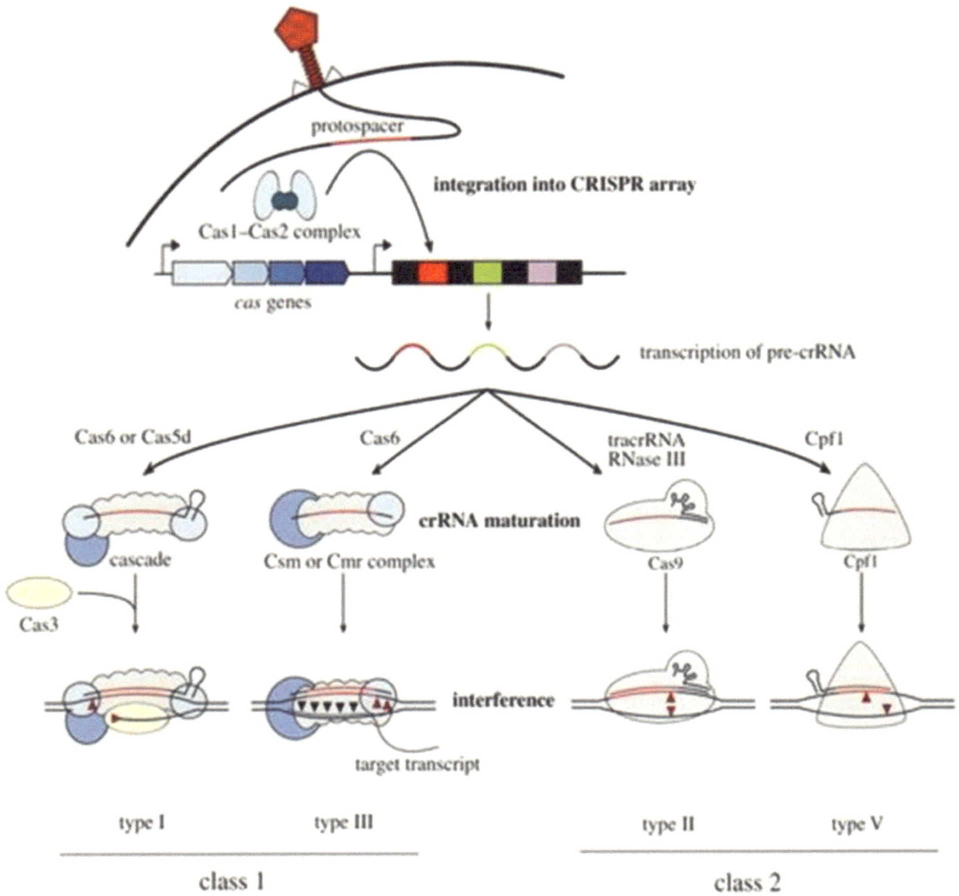

Figure 1.
Simplified model of the immunity mechanisms of class 1 and class 2 CRISPR-Cas systems. The CRISPR-Cas systems are composed of a cas operon (blue arrows) and a CRISPR array that comprises identical repeat sequences (black rectangles) that are interspersed by phage-derived spacers (colored rectangles). Upon phage infection, a sequence of the invading DNA (protospacer) is incorporated into the CRISPR array by the Cas1-Cas2 complex. The CRISPR array is then transcribed into a long precursor CRISPR RNA (pre-crRNA), which is further processed by Cas6 in type I and III systems (processing in type I-C CRISPR-Cas systems by Cas5d). In type II CRISPR-Cas systems, crRNA maturation requires tracrRNA, RNase III and Cas9, whereas in type V-A systems Cpf1 alone is sufficient for crRNA maturation. In the interference state of type I systems, Cascade is guided by crRNA to bind the foreign DNA in a sequence-specific manner and subsequently recruits Cas3 that degrades the displaced strand through its 3′–5′ exonucleolytic activity. Type III-A and type III-B CRISPR-Cas systems employ Csm and Cmr complexes, respectively, for cleavage of DNA (red triangles) and its transcripts (black triangles). A ribonucleoprotein complex consisting of Cas9 and a tracrRNA: crRNA duplex targets and cleaves invading DNA in type II CRISPR-Cas systems. The crRNA-guided effector protein Cpf1 is responsible for target degradation in type V systems. Red triangles represent the cleavage sites of the interference machinery (Courtesy: Ref. [4]).

Classification	Type I	Type II	Type III	Type V
Signature protein	Cas 3	Cas 9	Csm (III-A) or Cmr (III-B)	Cas 12a
Effector	Cascade	CrRNA	sgRNA	CrRNA
Cleavage product	SSBs	DSBs	SSBs	DSBs

Table 1.
Different types of CRISPR/Cas based on signature protein, effector, and cleavage product [3].

systems [1]. Due to the current problems with the vast diversity and complexity of the architecture of CRISPR/Cas systems, the classification is still challenging. Based on the presence of three signature genes, the classification is as follows:

1.1 Type I CRISPR/Cas systems

Typical type I loci contain the signature cas3 gene, which codes for helicase and DNase activities within a single large protein. The detailed sequence and structural comparisons have led to the recognition of many of these proteins in the RAMP superfamily including Cas5 and Cas6 families. Type I systems are currently divided into six subtypes, I-A to I-F, each of which has its own signature gene and distinct features of operon organization [2, 3].

1.2 Type II CRISPR/Cas systems

These contain cas9 as a signature gene encoding for a multidomain protein that combines all the functions of effector complexes and the target DNA cleavage and is essential for the maturation of the crRNA. These systems use cellular (not encoded within the CRISPR/Cas loci) RNase III and tracrRNA for the processing of pre-crRNA. Type II CRISPR-Cas systems are currently classified into three subtypes such as II-A, II-B, and II-C. Type II-A encompasses an additional signature gene csn2. Protein csn2 is found to be engaged in spacer integration. Type II-B systems belong to the Cas family of proteins with 5′-single-stranded DNA exonuclease activity. The recently proposed type II-C CRISPR-Cas systems possess only three protein-coding genes (cas1, cas2, and cas9) and are common in sequenced bacterial genomes (**Figure 1**) [2, 3].

1.3 Type III CRISPR/Cas systems

All type III systems possess the signature gene cas10 which encodes a multidomain protein containing a palm domain similar to that in cyclases and polymerases of the PolB family (**Table 1**) [2, 3].

2. Molecular characterization of CRISPR-Cas 12 and Cas 13

Initially, CRISPR-Cas 9 was found to nick the DNA along with the guide RNA Cas 12a belonging to class II Type VA system, derived from *Francisella novicida* bacterium possesses enormous ability to cleave DNA at multiple targets. Cas 12 an RNA-guided DNAse, is a T-rich PAM sequence making it different from Cas 9. The positively charged central channel of a nuclease (NUC) domain determines the trans cleavage activity of the target strand after studies find that mutations in the catalytic site of the RuvC domain of Cas12a in the bacterium *Acidaminococcus* sp. eliminate the same. CRISPR is classified into types I and II [5].

Type II is further divided into six types based on their structure and function. The Cas12a protein contains a RuvC endonuclease domain, which sequentially cleaves the non-targeting strand and the targeting strand to form DSBs (double-stranded base pairs). Compared to the CRISPR/Cas9 system, this system has several remarkable differences, including the signature protein, PAM sequence, and cleavage product [6]. CRISPR/Cas12a based sensing methods focus on fluorescence readout with reduced transduction efficiency as studies report a direct correlation between the catalysis systems with recognition elements (i.e., aptamers), thus greatly improving the working efficiency of the detection platform. Cas 13 consists of four subtypes and is involved in RNA interference activities. Off-target editing is critical to Cas 13 and requires significant attention in retrieving obstacles for protein analysis [7].

3. Mechanism of amplification strategy for nucleic acid and protein detection

CRISPR/Cas systems generally play a role as RNA-guided endonucleases (crRNA). The crRNA guides cas proteins to specific DNA sequences whereby the hybridization

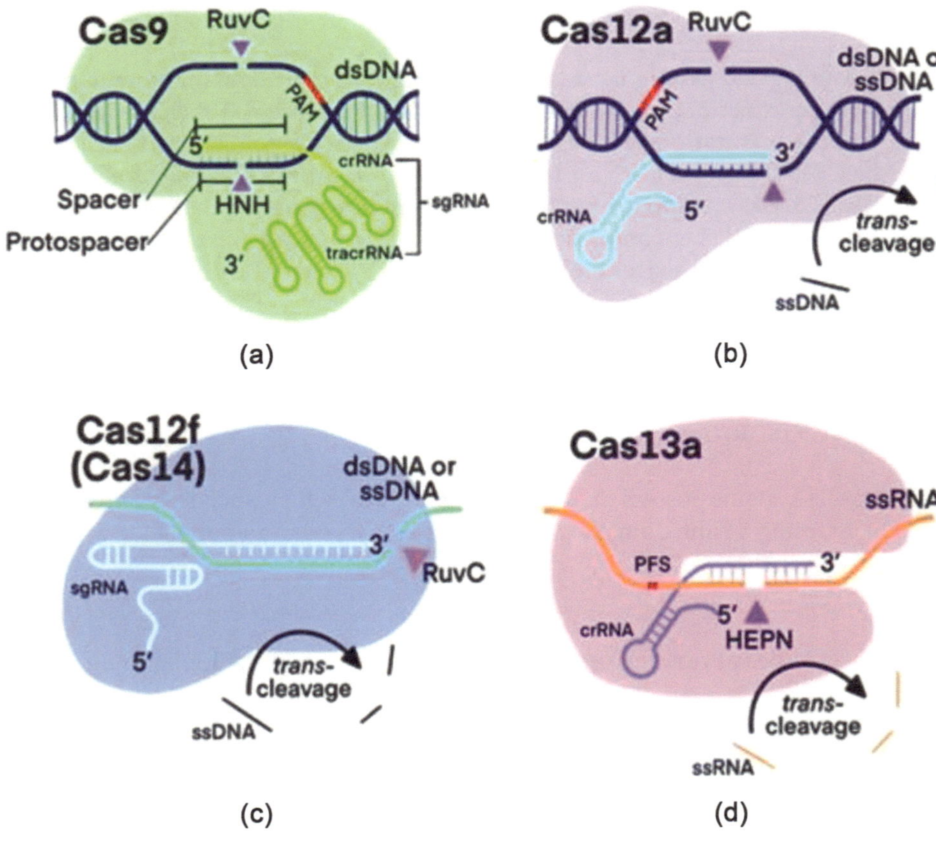

Figure 2.
Basic components of CRISPR/Cas9, Cas12a, Cas12f, and Cas13a pink triangle indicates cis-cleavage site [9].

leads to cas protein activation which later results in cleavage of DNA sequence [8]. **Figure 2** shows the components of Cas9, Cas12a, Cas12f, and Cas13a.

Cas9 is an endonuclease and the single-guide RNA (sgRNA) of CRISPR-Cas9 systems contains a hairpin-rich region that binds to Cas9 and a 20-nucleotide "spacer" region that binds with the complementary "protospacer" region in the target strand of a dsDNA duplex. Binding between the sgRNA and the DNA target brings Cas9 into close proximity to the target (**Figure 2**). The His-Asn-His (HNH) domain of Cas9

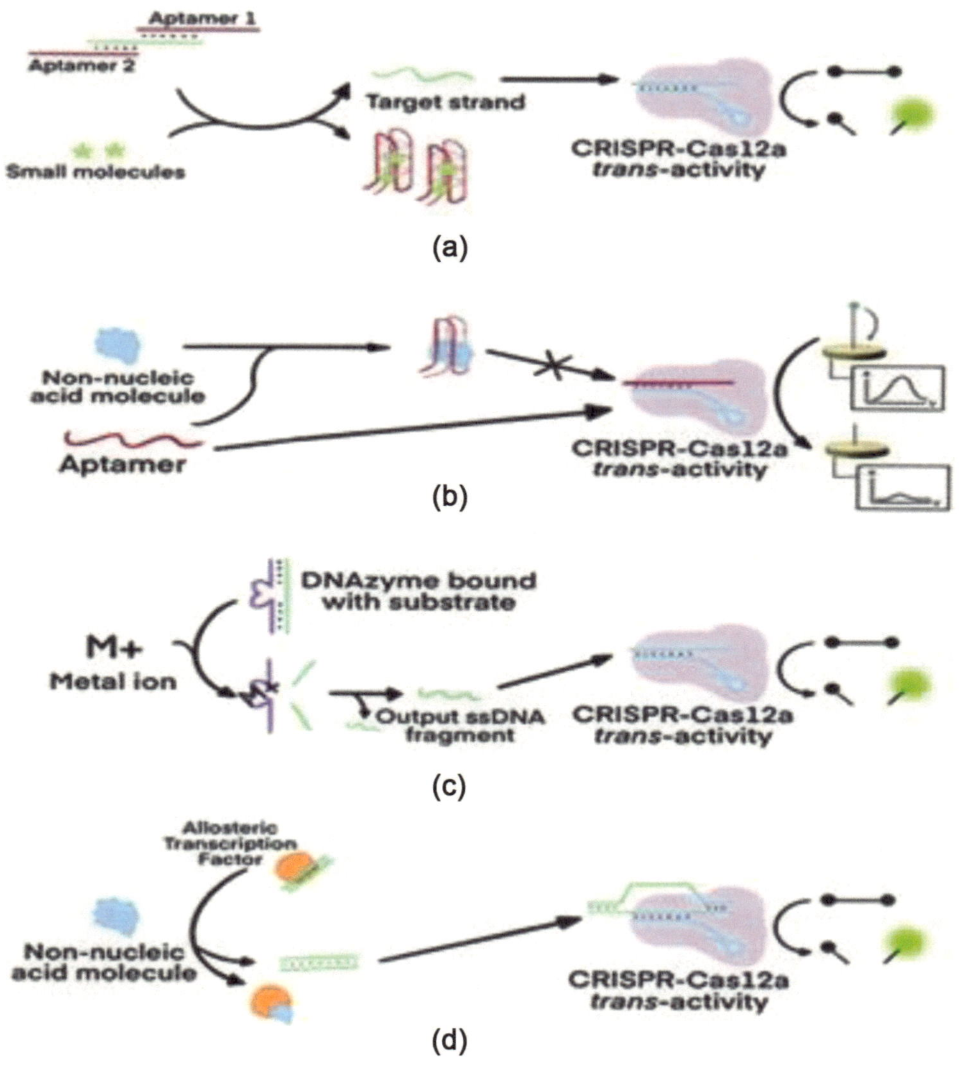

Figure 3.
Combining functional nucleic acids and molecular translators with CRISPR/Cas technology for detection of non-nucleic acids such as proteins. Adapted from Ref. [10]. Copyright 2019 Wiley-VCH Verlag GmbH & Co. KGaA, Weinheim. Adapted from Ref. [11]. Copyright 2020 American Chemical Society. Adapted from Ref. [12]. Copyright 2016 Springer Nature. (A) Two copies of an aptamer lock the activator, the target ssDNA complementary to crRNA. The activator is released when the aptamer binds to a small molecule (e.g., ATP), allowing it to hybridize with crRNA and activate CRISPR-Cas12a trans-activity. (B) The activation of CRISPR-Cas12a is prevented when the target molecule binds to its aptamer. CRISPR-Cas12a is activated by an unbound aptamer. (C) Metal ions serve as co-factor(s) for an RNA-cleaving DNAzyme to generate output ssDNA for CRISPR-Cas12a activation. (D) The binding of allosteric transcription factor (aTF) to the target molecule releases output dsDNA for CRISPRCas12a activation.

cleaves the strand that is complementary to sgRNA (target strand) and the RuvC domain of Cas9 cleaves the other strand of the dsDNA (non-target strand). Single-guide RNA (sgRNA) (**Figure 3**) [13].

Recent findings indicate that the cas12a proteins have both trans and cis cleavage activities on ssDNA regardless of the sequence. Notably, Cas12a is the first Cas protein to be identified whose ternary complex has been shown to have trans-ss DNA cleavage ability. Research shows that Cas12a may have acquired single-stranded DNA ability through evolution due to the abundance of viruses in the environment. Thus, gaining a significant role as a powerful and dominating weapon to eliminate invasion by foreign ss DNA. The well-characterized variants of cas12, cas12a, and cas12f, formerly known as cas14 lack the HNH domain but nevertheless, achieved the PAM dependant cleavage with RuvC domain alone. Recent findings have reported Cas13a also called C2Ca, an RNA-guided and RNA-targeting CRISPR effector from the class 2 type VI CRISPR system, was found to have the trans-cleavage activity on RNA. Additionally, the RuvC catalytic pocket of both C2c1 and Cas12a was responsible for the cleavage of both strands of targeted dsDNA [9].

4. Efficient sensing mechanism of CRISPR/Cas derived biosensors

Electrochemical biosensors register physical-chemical and biological change and possess high throughput of the biological recognition process. Depending on the type of biological recognition, sensors are classified into biocatalytic devices and affinity sensors. Biocatalytic sensors integrate enzymes and whole cells as recognition elements leading to exquisite specificity and a significant rise in the rate of reaction whereas affinity sensors make use of extreme selectivity and specificity for acquiring higher sensitivity. The electrochemical transducer responds to the binding event and converts the electrical response to an output that can be amplified, stored, and displayed [14]. Due to its signal-off architecture, these electrochemical sensors provide limited sensitivity and productivity. To overcome these limitations the CRISPR/Cas12a based electron sensing biosensors have been developed for non-nucleic acid targets. CRISPR/Cas12a-based immobilization-free electrochemical biosensors can detect small molecules and proteins by adjusting regions for target recognition in RCA components [15]. Transcription factors (TFs) assay seems to be path-breaking as it is found to be involved directly in many diseases including cancers. CRISPR/Cas 12a based biosensors for the detection of transcription factors have been developed. The biosensing mechanism is based on the interaction of TF's with double-stranded DNA activator eliminating Cas12a/crRNA from contacting and interacting with the 14 activators, thus inhibiting Cas12a activation. As a consequence of this strategy, the DNase activity of Cas12a was controlled and several TFs with well-defined binding sites could be quantified at the picomolar level with high precision [16, 17].

5. Implementation of CRISPR/Cas amplification strategy for protein detection

Recent findings report that the implementation of various nucleic acid amplification strategies led to improvements in analytical specificity and sensitivity and the development of point of care (POC) diagnostics. For example, the best-studied reaction is the amplification employing the Cas9 nickase (Cas9nAR) which when combined with polymerase and primers may substantially duplicate double-stranded DNA (dsDNA) without requiring heat cycling as does the polymerase chain reaction (PCR) [18].

5.1 iPCCA: isothermal proximity CRISPR/Cas 12a assay

In contrast to PCR, isothermal proximity assay seems to be an effective protein quantification assay for disease biomarkers and point of care diagnostics. Recent advances in the CRISPR/Cas technique specifically combining recombinase polymerase activity (RPA) and ssDNAse activity have led to the discovery of a series of isothermal assays for protein quantification. iPCCA relies on proximity binding for target recognition due to which it holds the potential for detecting non-nucleic acid targets such as proteins. However, isothermal amplification does not necessitate the use of advanced and sophisticated thermal cyclers and hence is more commonly used in biosensing [9].

5.2 Aptamer-based assay for femtomolar detection of proteins

The most widely used bioassay, ELISA (enzyme-linked immunosorbent assay) has revolutionized the ability to detect a wide variety of antigens. Complex chemical structure and restricted catalytic efficiency of HRP has a direct correlation with poor sensitivity in picomolar and nanomolar concentration. However, conventional ELISA is still not sensitive enough to detect ultralow concentrations of biomarkers for the early diagnosis of cancer, cardiovascular risk, neurological disorders, and infectious disease. CRISPR/Cas 13a based signal amplification strategy also called CLISA has been used to develop a 10 fold high-sensitive method for detecting low abundance. Recently, CRISPR/Cas13a has been recently demonstrated to have RNA-directed RNA cleavage ability. This RNA-guided trans-endonuclease activity is highly specific, being activated only when the target RNA has the perfect complementary sequence to the crRNA and is highly efficient. This potent signal amplification ability of CRISPR/Cas13a enables the development of direct RNA assays with a sensitivity down to the femtomolar level [19, 20].

5.3 CRISPR/Cas 12a controlled aptasensor for protein detection

Aptamer, a highly selective recognition element has been combined with various analytical techniques to increase the sensitivity of protein assay. Amongst these, an electrochemical technique using specific aptamers as recognition elements exhibits great promise in detecting protein duo to its attractive merits, such as high selectivity and sensitivity, the potential for miniaturization, and ease of integration with additional components [21]. Recent findings have demonstrated that the electrochemical aptasensor has been effectively used for the detection of thrombin in femtomolar concentration. It has been reported that once CRISPR RNA (crRNA)-directed Cas12a binds to a specific target DNA, the conserved RuvC nuclease domain in Cas12a will non-specifically cut single-stranded DNA (ssDNA). A homogeneous electrochemical aptasensor has been reported for sensitive and specific detection of thrombin by utilizing binding-induced DNA strand displacement strategy as the transduction element of thrombin and rolling circle amplification-regulated CRISPR/Cas12a for signal amplification. Importantly, this homogeneous electrochemical aptasensor can detect the femtomolar range of thrombin, and exhibited good specificity relative to other interfering blood-relevant proteins. The BIDSD-RCA-CRISPR/Cas12a is implemented in three steps, but this electrochemical aptasensor dispenses with the need for probe surface-immobilization procedure, simplifying the preparation process, and reducing the operating cost of the analysis. The strategy further could be applied to detect another disease-related protein biomarker in early diagnosis in the future [22].

6. Anti-CRISPRs: potential repressors of CRISPR/Cas

The struggle for life between bacteria and their infecting viruses (phages) has led to the development of numerous bacterial defense mechanisms and their phage-encoded opponents. Recently, anti-CRISPR proteins have been identified, which inhibits the CRISPR/Cas system. The mechanism by which anti-CRISPR proteins inhibit CRISPR/Cas provides an extensive set of valuable tools to both understand and manipulate CRISPR [21]. Several findings report that the growing number of anti-CRISPR families has a significant impact on CRISPR/Cas function and has been a driving force in the evolution of CRISPR-Cas. These Anti-CRISPR systems rely heavily on Aca proteins due to their extensive interaction with anti-CRISPRs and the presence of Aca genes has the potential to act as anti-anti CRISPR playing a vital role in CRISPR-based antibacterial technologies [22, 23]. Anti-CRISPR ranges from 50 to 150 amino acids with no sequence similarity. Recent finding demonstrated that phage carries atleast one anti-CRISPR gene to avoid elimination by competent hosts. The unique mechanism of anti-CRISPR results in sequence-specific transcriptional repression system. Type II anti-CRISPRs have more evident biotechnological uses, given the widespread usage of CRISPR-Cas9 genome editing tools. Their application could be critical for gene drive and gene therapy technologies [24].

7. Future perspectives of CRISPR-Cas technology

CRISPR/Cas based technology has a lot of potential as a tool for treating a range of medical conditions that have a genetic component, including cancer, hepatitis B, or even high cholesterol [25–27]. It is likely to be many years before CRISPR/Cas technology is used routinely in humans. CRISPR/Cas technology emerged as a versatile technology with wide application in the genome sequence editing, molecular studies of various gene functions, protein detection, gene therapy, and in the biomedical science as a diagnostic technology for detection of covid 19 like viral, bacterial, and various genetic disease [28]. Cancer is one of the fatal diseases that has severely threatened human life and caused a tremendous burden for society [29]. Early diagnosis of cancer is of great benefit to treat patients in early stages which leads to improve the survival rate of cancer patients. In body fluids detection of cancer related biomarkers is a critical kind of noninvasive technique for cancer diagnosis. Nevertheless, existing techniques of cancer biomarker detection always depends on a large-scale instruments and required sophisticated operation [30]. Clustered regularly interspaced short palindromic repeats and CRISPR-associated protein (CRISPR/Cas) based in vitro diagnosis can simplify the detection procedures and improve sensitivity and specificity, with great promise as the next-generation molecular diagnosis [31]. In the future, genome-wide screening for various genetic disorders, and cancer subtypes should be conducted to identify specific genetic and epigenetic targets for CRISPR technology to be most effective. The functionality of the identified mutations

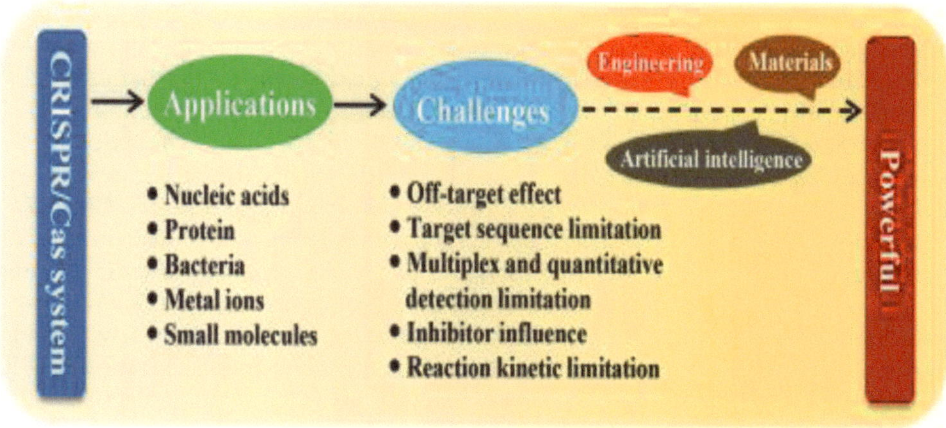

Figure 4.
Applications of CRISPR/Cas system in detection of various molecules [36].

and their related signaling pathways need to be thoroughly analyzed before they are manipulated for therapy with CRISPR technology [32]. More in vivo research on Cas9 epigenetic regulation is needed to better understand its impact on cancer epigenetics. The use of synthetic biology for Cas9 modulation can be further extended to create real-time predictive algorithms for specific metastatic pathways that update as epigenetic regulation progress and the cancer advances so that treatment can always be precisely one step ahead of cancer. Ongoing research has the potential to optimize and advance CRISPR technology, culminating in the clinical realization of its full potential for breast cancer diagnosis, modeling, and treatment [29, 33, 34]. In the future, CRISPR/Cas technology will be used as a unique promising technology to study the various genes for their function, for identification of mutations and their correction, this technology will be used in tumor angiogenesis research for cancer treatment [35], CRISPR technology also used for modification of genetic sequence to develop various organisms for the benefit of human and environmental protection. Much research is still focusing on its use in animal models or isolated human cells, with the aim to eventually use the technology to routinely treat diseases in humans (**Figure 4**).

8. Conclusion

From many years scientists have learned about genetics and gene function by studying the effects of alteration in DNA sequence. Artificially by making a change in a gene, either in a cell line or a whole organism, it is possible to study the effect of that change to understand what the function of that gene is. For a long period geneticists used chemicals or radiation to create mutations but this approach is not precise and specific and due to its randomness for several years scientists have been using 'gene targeting' to introduce changes in specific places in the genome, by deletion or insertion either whole genes or single bases. Conventional gene targeting has been very valuable for studying genes and genetics, however it takes a long time to create a mutation and is fairly expensive But the CRISPR/Cas9 system based technology currently stands out as the fastest, cheapest and most reliable system for 'editing' genes. In the last decade CRISPR/Cas is a genome editing technology that is creating a an atmosphere of excitement in the science world because of its faster, cheaper, promising, precise, sensitive and efficient and more accurate nature than previous conventional techniques of genome engineering and it has a wide range of potential applications. CRISPR/Cas technology have made it possible to edit the genomes of most cell types precisely and efficiently hence (CRISPR)/Cas9 system is a novel, versatile and easy-to-use tool to edit genomes irrespective of their complexity, with multiple and applications in almost all branches of life science, biomedicine and facilitating the elucidation of target gene function in biology and diseases. CRISPR/Cas technology able to detect various targets starting from nucleic acids to proteins. Incorporating CRISPR/Cas systems with numerous nucleic acid amplification strategies allows the generation of amplified detection signals, enrichment of low-abundance molecular targets, enhancements in analytical specificity and sensitivity, and development of point-of-care diagnostic techniques. It is concluded that the CRISPR/Cas systems in association with functional nucleic acids (FNAs) and molecular translators permits the detection of non-nucleic acid targets, like proteins, metal ions, and tiny molecules. Productive integrations of CRISPR technology with nucleic acid amplification techniques lead to sensitive and fast detection of Protein.

Acknowledgements

The authors would like to thank Dr. B.A. Mehere, Principal and Head of the Department of Biochemistry and Biotechnology, Dr. Ambedkar College, Deekshabhoomi, Nagpur, India, for providing research space and facility.

Conflict of interest

The authors declare no conflict of interest.

Author details

Rita Lakkakul and Pradip Hirapure*
Department of Biochemistry and Biotechnology, Dr. Ambedkar College, Deekshabhoomi, Nagpur, India

*Address all correspondence to: pradiphirapure@gmail.com

References

[1] Makarova KS, Haft DH, Barrangou R, Brouns SJ, Charpentier E, Horvath P, et al. Evolution and classification of the CRISPR–Cas systems. Nature Reviews Microbiology. 2011;**9**(6):467-477

[2] Wei C, Liu J, Yu Z, Zhang B, Gao G, Jiao R. TALEN or Cas9–rapid, efficient and specific choices for genome modifications. Journal of Genetics and Genomics. 2013;**40**(6):281-289

[3] Liu Z, Dong H, Cui Y, Cong L, Zhang D. Application of different types of CRISPR/Cas-based systems in bacteria. Microbial Cell Factories. 2020;**19**(1):1-4

[4] Hille F, Charpentier E. CRISPR-Cas: Biology, mechanisms, and relevance. Philosophical Transactions of the Royal Society of London. Series B, Biological Sciences. 2016;**371**(1707):20150496. DOI: 10.1098/rstb.2015.0496

[5] Makarova KS, Wolf YI, Iranzo J, Shmakov SA, Alkhnbashi OS, Brouns SJ, et al. Evolutionary classification of CRISPR–Cas systems: A burst of class 2 and derived variants. Nature Reviews Microbiology. 2020;**18**(2):67-83

[6] Yan F, Wang W, Zhang J. CRISPR-Cas12 and Cas13: The lesser known siblings of CRISPR-Cas9. Cell Biology and Toxicology. 2019;**35**(6):489-492

[7] O'Connell MR. Molecular mechanisms of RNA targeting by Cas13-containing type VI CRISPR–Cas systems. Journal of Molecular Biology. 2019;**431**(1):66-87

[8] Chen JS, Ma E, Harrington LB, Da Costa M, Tian X, Palefsky JM, et al. CRISPR-Cas12a target binding unleashes indiscriminate single-stranded DNase activity. Science. 2018;**360**(6387):436-439

[9] Feng W, Newbigging AM, Tao J, Cao Y, Peng H, Le C, et al. CRISPR technology incorporating amplification strategies: Molecular assays for nucleic acids, proteins, and small molecules. Chemical Science. 2021;**12**(13): 4683-4698

[10] Dai Y, Somoza RA, Wang L, Welter JF, Li Y, Caplan AI, et al. Exploring the trans-cleavage activity of CRISPR-Cas12a (cpf1) for the development of a universal electrochemical biosensor. Angewandte Chemie. 2019;**131**(48):17560-17566

[11] Xiong Y, Zhang J, Yang Z, Mou Q, Ma Y, Xiong Y, et al. Functional DNA regulated CRISPR-Cas12a sensors for point-of-care diagnostics of non-nucleic-acid targets. Journal of the American Chemical Society. 2019;**142**(1):207-213

[12] Liang M, Li Z, Wang W, Liu J, Liu L, Zhu G, et al. A CRISPR-Cas12a-derived biosensing platform for the highly sensitive detection of diverse small molecules. Nature Communications. 2019;**10**(1):1-9

[13] Nishimasu H, Ran FA, Hsu PD, Konermann S, Shehata SI, Dohmae N, et al. Crystal structure of Cas9 in complex with guide RNA and target DNA. Cell. 2014;**156**(5):935-949

[14] Li SY, Cheng QX, Liu JK, Nie XQ, Zhao GP, Wang J. CRISPR-Cas12a has both cis-and trans-cleavage activities on single-stranded DNA. Cell Research. 2018;**28**(4):491-493

[15] Ronkainen NJ, Halsall HB, HeinemanWR.Electrochemicalbiosensors. Chemical Society Reviews. 2010;**39**(5): 1747-1763

[16] Qing M, Chen SL, Sun Z, Fan Y, Luo HQ, Li NB. Universal and programmable rolling circle amplification-CRISPR/Cas12a-mediated immobilization-free electrochemical biosensor. Analytical Chemistry. 2021; **93**(20):7499-7507

[17] Li B, Xia A, Zhang S, Suo T, Ma Y, Huang H, et al. A CRISPR-derived biosensor for the sensitive detection of transcription factors based on the target-induced inhibition of Cas12a activation. Biosensors and Bioelectronics. 2021;**173**:112619

[18] Zhao X, Li S, Liu G, Wang Z, Yang Z, Zhang Q, et al. A versatile biosensing platform coupling CRISPR–Cas12a and aptamers for detection of diverse analytes. Science Bulletin. 2021;**66**(1):69-77

[19] Li Y, Mansour H, Watson CJ, Tang Y, MacNeil AJ, Li F. Amplified detection of nucleic acids and proteins using an isothermal proximity CRISPR Cas12a assay. Chemical Science. 2021; **12**(6):2133-2137

[20] Chen Q, Tian T, Xiong E, Wang P, Zhou X. CRISPR/Cas13a signal amplification linked immunosorbent assay for femtomolar protein detection. Analytical Chemistry. 2019;**92**(1):573-577

[21] Bounegru AV, Apetrei C. Carbonaceous nanomaterials employed in the development of electrochemical sensors based on screen-printing technique—A review. Catalysts. 2020;**10**(6):680

[22] Marotkar S, Hirapure P, Paranjape S, Upadhye V. Crispr/Cas9 technology for crop improvement: A new weapon for Indian agricultural threats. Plant Cell Biotechnology and Molecular Biology. 2020;**21**:1-9

[23] Li H, Li M, Yang Y, Wang F, Wang F, Li C. Aptamer-linked CRISPR/Cas12a-based immunoassay. Analytical Chemistry. 2021;**93**(6):3209-3216

[24] Pawluk A, Davidson AR, Maxwell KL. Anti-CRISPR: Discovery, mechanism and function. Nature Reviews Microbiology. 2018;**16**(1):12-17

[25] Zhang B. CRISPR/Cas gene therapy. Journal of Cellular Physiology. 2021;**236**(4):2459-2481

[26] Li H, Yang Y, Hong W, Huang M, Wu M, Zhao X. Applications of genome editing technology in the targeted therapy of human diseases: Mechanisms, advances and prospects. Signal Transduction and Targeted Therapy. 2020;**5**(1):1-23

[27] Sahel DK, Mittal A, Chitkara D. CRISPR/Cas system for genome editing: Progress and prospects as a therapeutic tool. Journal of Pharmacology and Experimental Therapeutics. 2019;**370**(3):725-735

[28] Tian X, Gu T, Patel S, Bode AM, Lee MH, Dong Z. CRISPR/Cas9—An evolving biological tool kit for cancer biology and oncology. NPJ Precision Oncology. 2019;**3**(1):1-8

[29] Alter HJ, Seeff LB. Recovery, persistence, and sequelae in hepatitis C virus infection: A perspective on long-term outcome. In: Seminars in Liver Disease. Vol. 20. New York, NY, USA: Thieme Medical Publishers, Inc.; 2000. pp. 17-36

[30] Gong S, Zhang S, Lu F, Pan W, Li N, Tang B. CRISPR/Cas-based in vitro diagnostic platforms for cancer biomarker detection. Analytical Chemistry. 2021;**93**:11899-11909

[31] Li Y, Li S, Wang J, Liu G. CRISPR/Cas systems towards next-generation

biosensing. Trends in Biotechnology. 2019;**37**(7):730-743

[32] Bondy-Denomy J, Garcia B, Strum S, Du M, Rollins MF, Hidalgo-Reyes Y, et al. Multiple mechanisms for CRISPR–Cas inhibition by anti-CRISPR proteins. Nature. 2015;**526**(7571):136-139

[33] Stanley SY, Borges AL, Chen KH, Swaney DL, Krogan NJ, Bondy-Denomy J, et al. Anti-CRISPR-associated proteins are crucial repressors of anti-CRISPR transcription. Cell. 2019;**178**(6):1452-1464

[34] Chen Y, Zhang Y. Application of the CRISPR/Cas9 system to drug resistance in breast cancer. Advanced Science. 2018;**5**(6):1700964

[35] Shah SZ, Zhao D, Hussain T, Sabir N, Yang L. Regulation of microRNAs-mediated autophagic flux: A new regulatory avenue for neurodegenerative diseases with focus on prion diseases. Frontiers in Aging Neuroscience. 2018;**10**:139

[36] Wu H, Chen X, Zhang M, Wang X, Chen Y, Qian C, et al. Versatile detection with CRISPR/Cas system from applications to challenges. TrAC Trends in Analytical Chemistry. Feb 2021;**135**:116150

Section 4

Gene Cloning for Nonribosomal Peptide Synthesis

Chapter 5

Nonribosomal Peptide Synthesis

Sadık Dincer, Hatice Aysun Mercimek Takci and Melis Sumengen Ozdenefe

Abstract

Nonribosomal peptides (NRPs) are a type of secondary metabolite with a wide range of pharmacological and biological activities including cytostatics, immunosuppressants or anticancer agents, antibiotics, pigments, siderophores, toxins. NRPs, unlike other proteins, are synthesized on huge nonribosomal peptide synthetase (NRPS) enzyme complexes that are not dependent on ribosomal machinery. Bacteria and fungi are the most common NRPs producers. Furthermore, the presence of these peptides has been confirmed in marine microbes. Nowadays, many of these peptides are used in the treatments of inflammatory, cancer, neurodegenerative disorders, and infectious disease for the development of new therapeutic agents. The structure, function, and synthesis of NRPs, as well as producer microorganisms and their several application areas, are covered in this chapter.

Keywords: biological activity, nonribosomal peptide, producer microorganisms, secondary metabolite, synthesis, structure

1. Introduction

Bioprocesses, which are consisted of a series of enzymatically catalyzed biochemical reactions in all living things, are necessary for survival. They have a high potential in terms of material synthesis, which has recently been performed by chemical techniques [1]. Furthermore, the advancement of heterologous production systems and genetic engineering techniques has resulted in pioneering initiatives to manufacture usable biomaterials [2]. These advancements also enabled the successful generation of primary and secondary metabolic pathway products in physiologically and genetically well-defined hosts, such as *Escherichia coli* and *Saccharomyces cerevisiae*, by precise manipulation of the related genes. In particular, heterologous molecular hosts have been used to successfully synthesize structurally varied secondary metabolites showing unique pharmacological action [1–3]. Nonribosomal peptides (NRPs) obtained by the most extensive, appealing, and useful actively-studied bioprocesses are included among these metabolites, which are important in the discovery and development of drugs and therapeutic reagents [1, 4].

NRPs are secondary metabolites that are synthesized outside of the ribosomal machinery and have a variety of properties such as cytostatics, immunosuppressants or anticancer agents, antibiotics, pigments, siderophores, toxins [3, 5, 6]. NRPs are typically produced by marine microorganisms and invertebrates, as well as soil-inhabiting microorganisms [5, 7, 8]. The majority of natural products produced by sponges, bryozoans, mollusks, and tunicates are members of the NRP or mixed polyketide–NRP families. Several of NRPs are being used in the development of new medicines for inflammatory, cancer, neurological diseases, and infectious disease nowadays [7].

Non-ribosomal peptide synthetases (NRPSs) enzymes are poly-functional mega-synthases that biosynthesize NRPs [7, 9, 10]. NRPSs, multi-modular enzyme or enzyme complexes from common bacteria, less common eukarya, and rare archaea, are capable of producing a wide range of natural pharmaceutical products (Bacitracin, antibiotics for skin infections; Bleomycin antitumor; Cyclosporin, antifungal, and immunosuppressive drugs; Daptomycin, antibiotics) [5, 7, 11]. NRPSs use both proteinogenic and nonproteinogenic amino acids (not encoded by DNA) as building blocks for the growing peptide chain [1, 7, 11, 12]. They catalyze multiple biosynthetic processes, each of which is responsible for particular one amino acid elongation on the growing peptide chain [10]. This chapter looks at the structure, function, and synthesis of NRPs, as well as producer microorganisms and their various applications.

2. Synthesis, structure, and function of nonribosomal peptide (NRP)

NRPSs are responsible for nonribosomal peptide (NRP) synthesis. These are large multi-enzyme complexes that are modularly organized and serve as biosynthetic machinery and templates [5, 11–14]. For example, a single NRPS of 1.6 MDa synthesizes the Cyclosporine A (7). In fungal systems, such as in the case of cyclosporine A (7), a single NRPS synthesizes peptides, whereas bacteria frequently use numerous NRPSs with genes grouped in an operon. NRPSs have a modular structure [14, 15].

In a genome mining research of 2699 genomes, Wang et al. found that more than half of the NRPS enzymes were non-modular NRPS enzymes [16]. Nonmodular NRPS enzymes can be found in siderophore biosynthesis pathways, such as EntE and VibH in enterobactin and VibE in vibriobactin, or as a standalone peptidyl carrier protein, such as BlmI from the bleomycin gene cluster. NRPS enzymes are found frequently in bacteria, less frequently in eukaryotes, and infrequently in archaea. Actinobacteria, Cyanobacteria, Firmicutes, and Proteobacteria were the phyla with the greatest number of these enzymes in the bacterial domain. There was a correlation between genome size and the number of NRPS clusters [5, 17].

A module is a part of the NRPS polypeptide chain that is in charge of integrating one amino acid into the final product. Modules can further be separated into domains (**Figure 1**), which represent enzyme units that catalyze distinct steps of NRP synthesis. On the protein level, domains are defined by distinctive, greatly conserved order of patterns known as "core motifs." In certain instances, biochemical and structural data were used to confirm the involvement of greatly conserved residues in domain function (**Table 1**) [14].

There are three domains in a module. These are 1) the adenylation (A) domain, 2) the peptidyl carrier protein (PCP) or thiolation (T) domain, and 3) the condensation (C) domain, all of which are responsible for the synthesis of NRPs. A module may include additional tailoring or altering domains incorporating epimerization (E), methylation and oxidation domains or a heterocyclization (Cy) domain in place of a C-domain. Finally, most NRPS termination modules have a TE-domain, which is in charge of releasing linear, cyclic, or branching cyclic peptides [5, 9–11, 15, 18–21].

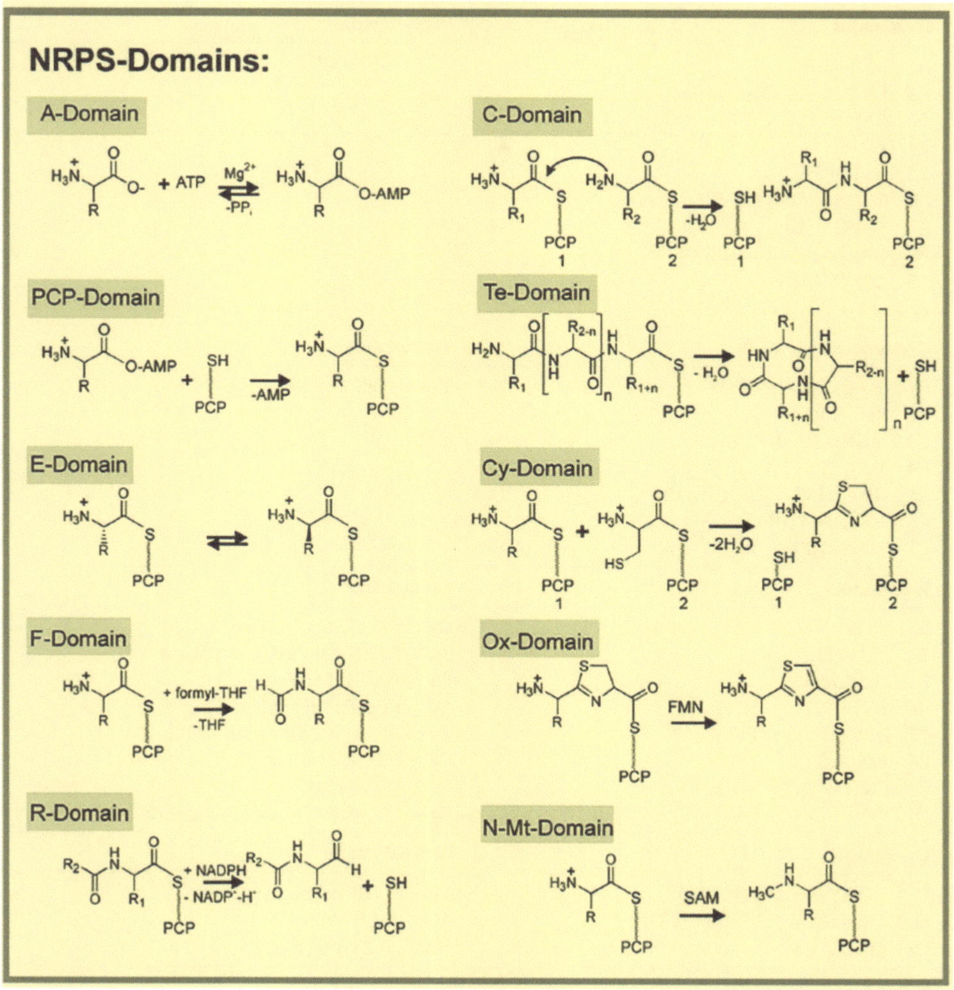

Figure 1.
Catalyzed reactions by various NRPS-domains [14].

The order of the modules is frequently aligned with the sequences of the resulting peptides. NRP synthesis begins at the N-terminus and ends at the C-terminus, yielding peptides that are typically 3–15 amino acids long. The released peptides contain amino acids, that is, imino acids or ornithine and their structures are linear, cyclic-macrocyclic, branched-cyclic, branched-macrocyclic, dimers or trimers of identical structural elements [5].

The A-domain is responsible for the first step in biosynthesis, which involves recognizing and activating the amino acid substrate via adenylation with Mg-ATP, resulting in an aminoacyl adenylated intermediate. Around 550 amino acids make up domain A. It has 10 amino acid residues that serve as NRPS enzyme "codons" and are essential for substrate specificity. The D and L forms of the 20 amino acids used in ribosomal protein synthesis, as well as non-proteinogenic amino acids like imino acids, ornithine, and hydroxy acids like β-butyric acids and α-aminoadipic, are substrates recognized by the A-domain. The PCP-domain, which consists of about 80 amino acids and covalently attaches the activated amino acid to their cofactor 4′-phosphopantetheine (PP)

A-domains	**PCP-domains**
A1 L(TS)YxEL	T LGG(DH)SL
A2 LKAGxAYL(VL)P(LI)D	
A3 LAYxxYTSG(ST)TGxPKG	
A4 FDxS	
A5 NxYGPTE	
A6 GELxJGx(VL)ARGYL	
A7 Y(RK)TGDL	
A8 GRxPxQVKIRGxRIELGEIE	
A9 LPxYM(IV)P	
A10 NGK(VL)DR	
C-domains	**Te-domains**
C1 SxAQxR(LM)(WY)xL	Te GxSxG
C2 RHExLRTxF	
C3 MHHxISDG(WV)S	
C4 YxD(FY)AVW	
C5 (IV)GxFVNT(QL)(CA)xR	
C6 HN)QD(YD)PFE	
C7 RDxSRNPL	
E-domains	**Cy-domains**
E1 PIQxWF	Cy1 FPL(TS)xxQxAYxxGR
E2 HHxISDG(WV)S	Cy2 RHx(IM)L(PAL)x(ND)GxQ
E3 DxLLxAxG	C3 D(NLI)xDxxS
E4 EGHGRE	Cy3 LPxxPxLPLxxxP
E5 RTVGWFTxxYP(YV)PFE	Cy4 (TS)(PA)3x(LAF)6x(IVT)LxxW
E6 PxxGxGYG	Cy5 (GA)DFTxLxLL
E7 FNYLG(QR)	Cy6 PVVFTSxL
	Cy7 (ST)(QR)TPQVx(LI)D13xWD
Ox-domains	**N-Mt-domains**
Ox1 KYxYxSxGxxY(PG)VQ	M1 VL(DE)xGxGxG
Ox2 GxxxG(LV)xxGxYYY(HD)P	M2 NELSxYRYxAV
Ox3 IxxxYG	M3 VExSxARQxGxLD
R-domains	
R1 V(L)(L)TG(A)TG(F)(L)GxxLL	
R2 Vx(L)(L)VR(A)	
R3 GPL(G)x(P)x(L)GL	
R4 V(Y)PYxYLxx(P)NVxxT	
R5 GYxxSKW(A)(A)E	
R6 R(P)G	
R7 YxxxxG(LF)LxxP	

Table 1.
NPRS-domains' core-motifs [14].

arm via a thioester bond, completes the second step. And also, the active substrate and elongation intermediates are transferred to the C-domain via this domain. In the last step, C-domain, which contains approximately 450 amino acids, catalyzes the formation of peptide bonds between the carboxyl group of the incipiently synthesized peptide and the amino acid transported by the side module [5, 22]. Furthermore, this domain allows the expanding chain to be translocated to the next module. Following this step, the linear intermediate peptide is liberated in bacteria via internal cyclization or hydrolysis with the help of the Thioesterase (TE) domain. On the other hand, it appears less commonly in fungi's NRPSs. Fungi use a variety of ways to release chains. The first is a thioesterase NADP(H)-dependent reductase domain (R), which catalyzes NADPH reduction to create an aldehyde and the second is a terminal C domain, which catalyzes release by forming intermolecular or intramolecular amide bonds. By N-, C-, and O-methylation, halogenation, acylation, hydroxylation, glycosylation, or heterocyclic ring formation, the primary product of this synthesis can be changed post-synthetically to reach its mature form by additional tailoring enzymes that are not part of the NRPS. The structural diversity of NRPs is formed in part by these enzymes and their reactions [5].

Because of their extensive multidomain organization, NRPS genes are easier to identify using recent genome mining technologies, and they are also relatively easy to detect. Secondary metabolites production genes are frequently found in bacterial and fungal gene clusters. The clusters' core is thought to be NRPS genes. Nevertheless, they are linked to genes involved in building blocks synthesis, product ornamentation, self-resistance, and peptide export. For the purpose of analyzing and in silico exploration of NRPS pathways, advanced genome sequencing techniques have made genome mining methodologies available, which are assisted by a variety of bioinformatics tools, such as AntiSMASH, PRISM, and SMURF [23].

Nowadays, known NRP structures are divided into various categories, each with its own structural characteristics. Lipocyclopeptides with varied linkage patterns, such as fengycin, iturin, surfactin, and head-to-tail-cyclized peptides of varying ring sizes, such as cyclosporine, gramicidin S, maybe the largest group. There are also a lot of linear peptide configurations. They include tripeptides (such as sevadicin and bialaphos) as well as 20-mer peptides (e.g., alamethicin, peptaibols). The current highest size limit for NRPs is syringopeptin 25A, which has 25 amino acids (syringopeptin 25A). Tailoring enzymes modify the structure of some NRPs. The most structurally complicated molecules are probably bleomycins, ergopeptides, glycopeptide antibiotics, and β-lactams [23].

Figure 2 shows some NRPs with diverse structures and a wide spectrum of activities. ACV-tripeptide (6), for example, is a precursor to antibiotics of the penicillin and cephalosporin families. Gramicidin S (4), tyrocidine A (1), and vancomycin (5) are three other antibiotic-active substances. Cyclosporin A (7), an immunosuppressive drug, is used in the post-transplantation care of patients. Cancer is treated with cytostatic agents, such as bleomycin A2 (8) and epothilone (9). Enterobactin (10), bacillibactin (11), mixochelin A (12), yersiniabactin (13), and vibriobactin (14) are examples of iron chelating agents. These compounds, known as siderophores, are created in iron-deficient environments to provide bacteria with an iron source. **Figure 2** also depicts the structures of pigments like indigodin (15), toxins like thaxtomin A (17), and peptides with uncertain functions like anabaenopeptilide 90-A (18) [14].

NRPs have a number of structural characteristics that distinguish them from ribosomal peptides. For example, non-proteinogenic amino acids, such as ornithine in 1, 2, and 4, hydroxyphenyl or dihydroxyphenyl-glycine in 5 and (4R)-4-[(E)-2-butenyl]-4-methyl-L. -threonine (Bmt) in 7, are included. Furthermore, the structures are frequently macrocyclic (1), branched macrocyclic (2), or dimers of two (4) or trimers of three (10, 11) structural components. Smaller heterocyclic rings, such as thiazole in 9, thiazoline in 13, or oxazoline in 14, are common structural properties of nonribosomal peptides. In addition, fatty acids (3), glycosylations (5), N-methylations (7), and N-formylations (18) may also be present, as well as the addition of propionate units (8) or acetate [14].

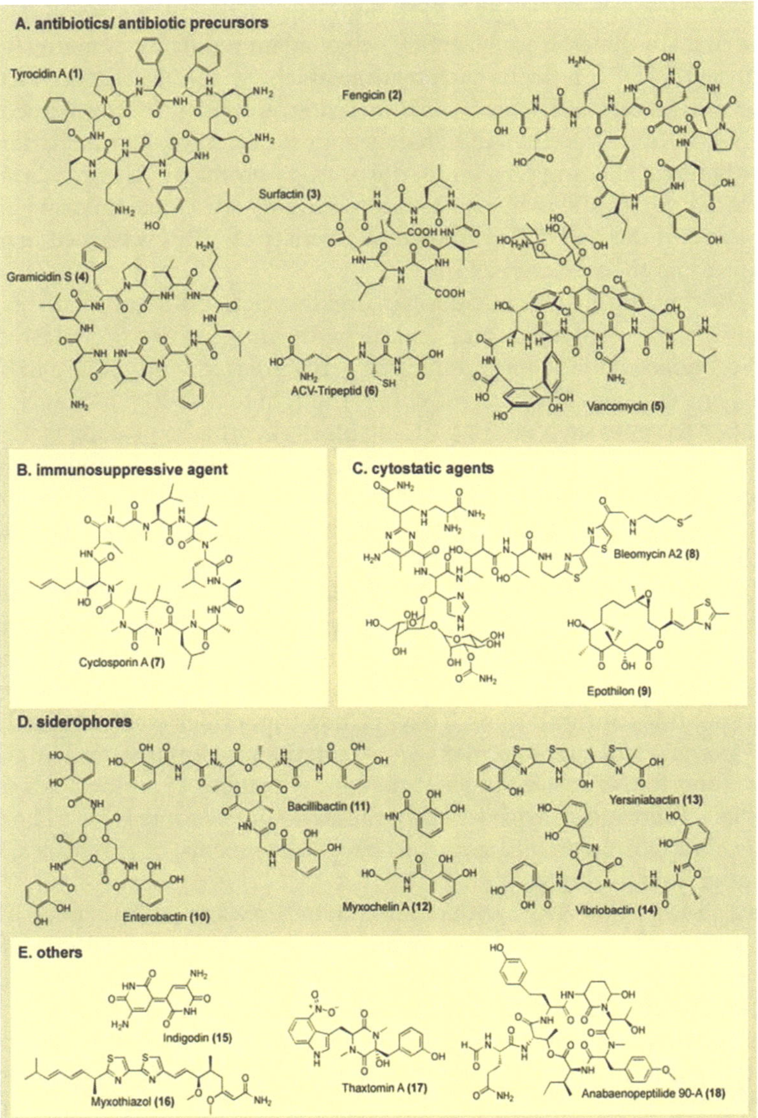

Figure 2.
Some NRPs with structural diversity [14].

3. Overview of producer microorganisms for NRP

NRPs are typically produced by marine microorganisms, soil-inhabiting microorganisms, including *Actinomycetes*, *Bacilli,* and eukaryotic filamentous fungus, and invertebrates, such as sponges, bryozoans, mollusks, and tunicates [5, 7, 11, 13, 24]. Many pharmacologically active NRPs have been effectively generated in heterologous hosts, such as *Bacillus subtilis*, *Escherichia coli*, *Saccharomyces cerevisiae,* and *Streptomyces* sp. [2]. Bacteria and fungi are the primary producers of NRPS-based metabolites. Except for bacteria and fungus, NRPS Ebony from *Drosophila melanogaster* ("fruit fly") and nemamide synthetase from the worm *Caenorhabditis elegans* have been confirmed. The distribution and occurrence of NRPS pathways and products have been discovered, thanks to screening efforts and genome sequencing projects followed by bioinformatics research. NRPS enzymes are found frequently in bacteria, less frequently in eukaryotes, and infrequently in archaea. The phylum Actinobacteria (*Mycobacterium*, *Streptomyces*), Firmicutes (*Bacillus*, *Staphylococcus,* and *Streptococcus*), and the alpha-/beta-/gama-Proteobacteria classes (*Burkholderia*, *Escherichia*, *Erwinia, Photorhabdus, Pseudomonas*, *Salmonella*, *Serratia*, *Vibrio*, and *Yersinia*) are the most important contributors among bacteria. Nonetheless, in recent years, the phylum Cyanobacteria (*Microcystis*, *Planktothrix*, *Anabaena*, *Oscillatoria,* and *Nostoc*) and the teta-Proteobacteria (*Myxobacterium*) class have received greater attention [5, 22, 23]. NPRS genes are found predominantly in the Ascomycota (*Tolypocladium*, *Fusarium*, *Penicillium*, *Acremonium*, *Claviceps,* and *Trichoderma*) and marginally in the Basidiomycota (*Ustilago*) phylum. NRPS biosynthesis investigations in fungus are less investigated than in bacteria due to greater genome sizes, the existence of scattered introns in gene clusters, and a less established molecular biology toolbox [23].

4. Application areas of NRPs

Novel peptide products' biological functions are strictly associated with their chemical structure, which is constrained by a peptide sequence that ensures specific interaction with a specific molecular target. Chemical alterations, such as the incorporation of fatty acid chains, D-amino acids, glycosylated amino acids, and heterocyclic rings, as well as cyclization or oxidative cross-linking of side chains, add a lot to these unique interactions. Bacitracin, fengycin, pristinamycin, surfactin, tyrocidine, and vancomycin are examples of novel peptides with antibacterial and antifungal properties [25].

When the ribosomal code was deciphered in the 1960s, Tatum and coworkers discovered that ribosomes had no effect on cell-based tyrocidine production [23, 26]. The first NRPs agent is tyrocidine, a cyclic decapeptide that is biosynthesized outside of the *Bacillus brevis* ribosome. Researchers discovered that ribosome targeting antibiotics had no effect on tyrocidine production. They also discovered that *B. brevis* can synthesize gramicidin S, a cyclic decapeptide, without the use of tRNA molecules or aminoacyl-tRNA synthetases [13, 27]. Nobel Prize Laureate Fritz Lipmann and Søren Laland contributed to present essential biochemical activity insights into NRPSs, including specific ATP-dependent activation of amino acids, thioester-mediated 4′-phosphopantetheine (Ppant) binding of activated amino acids, and the directionality of the peptide synthesis and have given acceleration to the production of NRPS-based metabolites synthesized by a mechanism distinct from protein synthesis. The NRPs and NRPSs were discovered as a result of these findings associated with the synthesis of tyrocidine and gramicidin S peptides. Surprisingly, the majority of studies investigating nonribosomal NRPS-based metabolites have focused on antibacterial and antifungal action [23]. NRPS-based metabolites with antimalarial, antimicrobial, antiparasitic, antiviral, animal growth promoters, cytostatic, immunosuppressive, and natural insecticides properties are currently available on the market, and several are being studied in clinical research [28]. **Table 2** presents a summary of commercialized NRPs-based medications with antibacterial activity.

As demonstrated in **Table 2**, systemic and topical antibacterials are the most often used NRPs-based drugs, accounting for billions of dollars in the chemical and pharmaceutical industry sales. **Table 3** lists their other applications, which include

Compound	Biosynthetic class of agent	Source	Disease/Molecular target
Bacitracin	Cyclic peptide	*Bacillus subtilis*	Antibiotic/dephosphorylation of C55-isoprenyl pyrophosphate
Bleomycin	Hybrid peptide	*Streptomyces verticillus*	Antibiotic/inhibition of DNA synthesis
Capreomycin	Cyclicpeptide	*Streptomyces capreolus*	Antibiotic/protein synthesis inhibitor
Carbapenems	Synthetic thienamycin	*Streptomyces cattleya*	Antibacterial (multidrug resistant)/ bacterial cell-wall biosynthesis (peptidoglycan;β-lactamase inhibition)
Cephalosporin	*β*-lactam	*Acremonium chrysogenum*	Antibiotic/Alters bacterial outer membrane
Chlorampheniol	Synthetic;further derivatives: thiamphenicol [c], florfenicol	*Streptomyces venezuelae*	Antibacterial/inhibition of ribosomal protein synthesis
Colistin (Polymyxin E)	—	*Paenibacillus polymyxa var. colistinus*	Antibacterial/binding to lipopolysaccharide (outer membrane), interaction with the cytoplasmic membrane
Dalbavancin	Semisynthetic teicoplanin derivative	—	Antibacterial (Gram-positive)/ membrane anchoring; disruption of cell membrane and inhibition of bacterial cell wall biosynthesis
Daptomycin	Lipopeptide	*Streptomyces roseosporus*	Antibiotic (Gram-positive)/disrupts the cell membrane
Gramicidin	Linear pentadecapeptide	*Bacillus brevis*	Antibiotic/ion-channel formation, increasing the permeability of the membrane
Lincomycin	—	*Streptomyces lincolnensis*	Antibacterial (patients allergic to penicillin) inhibition of the ribosomal protein synthesis (50S-subunit, dissociation of peptidyl-tRNA from the ribosome)
Monobactams	—	*Chromobacterium violaceum*	Antibacterial (Gram-negative)/ bacterial cell-wall biosynthesis
Oritavancin	—	Semi synthetic	Antibiotic/disrupts the cell membrane
Polymyxin B	Polypeptides	*Bacillus polymxya*	Antibacterial (Gram-negative)/ binding to lipopolysaccharide (outer membrane), interaction with cytoplasmic membrane
Pristinamycin	Depsipeptide	*Streptomyces pristinaespiralis*	Antibacterial (Gram-positive)/ ribosomal biosynthesis (50S-subunit, peptidyl transfer, and elongation of protein synthesis)
Teicoplanin	Glycopeptide	*Actinoplanes teichomyceticus*	Antibiotic/inhibit cell wall synthesis
Telavancin	—	*Amycolatopsis orientalis*	Antibacterial (Gram-positive) disruption of cell membrane and inhibition of bacterial cell-wall biosynthesis

Compound	Biosynthetic class of agent	Source	Disease/Molecular target
Tyrothricin	—	*Bacillus brevis*	Antibacterial (Gram-positive)/ disruption of cell membrane
Vancomycin	Glycopeptide	*Amycolatopsis orientalis*	Antibiotic/inhibit cell wall synthesis
Virginiamycin	—	*Streptomyces virginiae*	Antibacterial/ribosomal biosynthesis (50S-subunit, peptidyl transfer, and elongation of protein synthesis)

Table 2.
Overview of NRPs-based drugs [7, 23].

Agent	Origin	Properties and area of application
Actinomycin D (Dactinomycin)	*Actinomyces antibioticus*, *Streptomyces chrysomallus*	Antitumor/DNA intercalator, inhibition of transcription
Bialaphos	*Streptomyces hygroscopicus*, *Streptomyces viridochromogenes*	Herbicide/tripeptide prodrug, inhibitor of glutamine synthetase
Bleomycin A2, B2	*Streptomyces verticillus*	Antitumor/metal-dependent oxidative cleavage of DNA in presence of molecular oxygen
Capreomycin	*Streptomyces capreolus*	Antituberculous/ inhibition of the ribosomal protein synthesis (16S and 23S-rRNA)
Carfilzomib	Synthetic derivative of epoxomycin (*Actinomyces* sp.)	Anticancer/proteasome inhibitor
Caspofungin	*Glarea lozoyensis*, semisynthetic from pneumocandin; further derivatives: micafungin/ anidulafungin	Antifungal (candidiasis, aspergillosis) fungal cell-wall integrity ((1-3)-β-D-glucan synthase)
Cyclosporine A	*Tolypocladium inflatum*	Immunosuppressant/cyclophilin binding, inhibition of IL-2 expression (inhibition of T-cell activation)
Emodepside	*Mycelia sterilia* (F); semisynthetic from PF1022A	Anthelmintic/Slo-1 receptor (K+ channel)
Enduracidin (Enramycin)	*Streptomyces fungicidicus*	Antibacterial, food additive/inhibition of MurG (essential for cell-wall biosynthesis in Gram positive bacteria), inhibition of the transglycosylation step of peptidoglycan biosynthesis
Enniatins (fusafungine)	*Fusarium lateritium*, *Fusarium scirpi*, *Fusarium* sp.	Antibacterial (topical), antifungal, anti-inflammatory/ ionophore ($NH4^+$) membrane depolarization
Ergometrine (ergonovine)	*Claviceps purpurea*	Obstetrics/interaction with a-adrenergic, dopaminergic and serotonin receptors
Ergotamine	*Claviceps purpurea*	Migraine vasoconstrictive (5-HT1B receptor, but also dopamine and noradrenaline receptors)
Romidepsin	*Chromobacterium violaceum*	Antitumor/histone deacetylase inhibitor (inducing apoptosis)
Trabectedin	Bacterial symbiont of *Ecteinascidia turbinata* (sea squirt)	Antitumor (antiproliferative, treatment of soft tissue sarcoma) DNA binder, blocks binding of transcription factors

Table 3.
Marketed-NRPs agents [23].

anticancer agents, antifungals, animal feed additives, immunosuppressants (cyclosporine), obstetrics (ergometrine), and pain management (ergotamine).

In the medical field, NRP-based marketed drugs, such as Cyclosporin A and Bleomycin A2, have high selling prices. The cost of these medicines is $107 for 25 mg of Cyclosporine A (98% purity) obtained from *T. inflatum* and $847 for 20 mg of Bleomycin A2 (70% purity) isolated from *S. verticillus*, according to Sigma Chemical Company [5].

The 70% discovery of NRPs with antibacterial, antiviral, cytostatic, immunosuppressive, antimalarial, antiparasitic, animal growth promoters, and natural insecticides activity is mostly attributed to marine organisms [13]. NRPs obtained from marine organisms (sponges, tunicates, and their associated phyla, such as Acidobacteria, Actinobacteria, Bacteriodetes, Chloroflexi, Cyanobacteria, Nitrospira, Planctomycetes, Poribacteria, Proteobacteria, Verrucomicrobia, and Archaea) have excellent binding properties, low off-target toxicity, and high stability and these properties make them a promising molecule for the development of new therapeutics pharmacologically active in many clinical searches. **Table 4** shows the chemical structure and source of various NRPs isolated from marine sponges and tunicates.

NRPs agents	Chemical class	Origin	Target
Miraziridine A	Linear pentapeptide	*Theonella* aff. *mirabilis*	Cancer/inhibit protease cathepsin B
Haligramides A-B	Cyclic hexapeptides	*Haliclona nigra*	Cancer/A-549 (lung), HCT-15 (colon), SF-539 (CNS), SNB-19 (CNS)
Prepatellamide A	Cyclic peptide	*Lissoclinum patella*	Cancer/P388 murine leukemia cell lines
Tamandarins A-B	Depsipeptides	*Didemnid ascidian*	Cancer/pancreatic carcinoma BX-PC3, prostatic cancer DU-145, head and neck carcinoma UMSCC10b
Microsclerodermins F–I	Cyclic peptides	*Microscleroderma* sp.	Cancer/HCT-116 cell line
Wainunuamide	Cyclic hexapeptide	*Stylotella aurantium*	Cancer/A2780 ovarian, K562 leukemia cancer cells
Leucamide A	Cyclic hexapeptide	*Leucetta microraphis*	Cancer/Tumor cell lines HM02, HepG2, Huh7
Axinellin C	Cyclic octapeptide	*Stylotella aurantium*	Cancer/A2780 ovarian, K562 leukemia cancer cells
Milnamide D	Linearpeptide	*Cymbastela* sp.	Cancer/HCT-116 cells
Kapakahines E–G	—	*Cribrochalina olemda*	Cancer/P388 murine leukemia cells
Didmolamides A-B	Cyclic hexapeptides	*Didemnum molle*	Cancer Tumor cell lines (A549, HT29 and MEL28)
Bistratamides E–J	Cyclic hexapeptides	*Lissoclinum bistratum*	Cancer/Human colon tumor (HCT-116) cell line
Milnamide C	—	*Auletta* sp.	Cancer/MDA-MB-435cancer cells
Scleritodermin A	Cyclic peptide	*Scleritoderma nodosum*	Cancer

NRPs agents	Chemical class	Origin	Target
Microcionamids A-B	—	*Clathria abietina*	Cancer/Human breast tumor cell lines MCF-7 and SKBR-3
Kendarimide A	Linear peptide	*Haliclona* sp.	Cancer/KB-C2 cells
Phakellistatin 14	Cyclo heptapeptide	*Phakellia* sp.	Cancer/Murine lymphocytic leukemia P388 cell line
Polytheonamides A-B	Polypeptides	*Theonella swinhoei*	Cancer/P388 murine leukemia cells
Neopetrosiamides A-B	Tricyclic peptides	*Neopetrosia* sp.	Cancer
Seragamides A–F	Depsipeptides	*Suberites japonicus*	Cancer
Theopapuamide	Cyclic depsipeptide	*Theonella swinhoei*	Cancer/CEM-TART, HCT-116 cell lines
Azumamide A-E	Cyclo tetrapeptides	*Mycale izuensis*	Cancer
Callyaerin G	Cyclic peptide	*Callyspongia aerizusa*	Cancer/Mouselymphoma cell line (L5178Y) and HeLa cells
Stylopeptide 2	Cyclo decapeptide	*Stylotella* sp.	Cancer/BT-549 and HS578T breast cancer cell lines
Ciliatamides A-C	Lipopeptides	*Aaptos ciliate*	Cancer/HeLa cells
Diazonamides C–E	Macrocyclic peptides	*Diazona* sp.	Cancer/Human tumor cell lines (A549, HT29, MDA-MB231)
Rolloamide A-B	Cyclic heptapeptides	*Eurypon laughlini*	Cancer
Euryjanicin A	Cycloheptapeptide	*Prosuberites laughlini*	Cancer
Callyaerin A–F and H	Cyclic peptides	*Callyspongia aerizusa*	Cancer/L5178Y cell line
Papuamides E-F	Depsipeptides	*Melophlus* sp.	Cancer/Brine shrimp
Stylissamide X	Octapeptide	*Stylissa* sp.	Cancer/HeLa cells
Gombamide A	Hexapeptide	*Clathria gombawuiensis*	Cancer/K562 and A549 cell lines
Microspinosamide	Cyclic depsipeptide	*Sidonops microspinosa*	HIV
Neamphamide A	Cyclic depsipeptide	*Neamphius huxleyi*	HIV
Mirabamides A-D	Cyclic depsipeptide	*Siliquariaspongia mirabilis*	HIV
Homophymine A	Cyclodepsipeptide	*Homophymia* sp.	HIV/PBMC cell line
Celebeside A-C	Depsipeptides	*Siliquariaspongia mirabilis*	HIV/Colon carcinoma (HCT-116) cells
Theopapuamides B–D, Mutremdamide A, Koshikamides C-H	Cyclic depsipeptide	*Theonella* sp.	HIV
Ceratospongamide	Cyclic heptapeptide	*Sigmadocia symbiotica*	Inflammation
Halipeptin A-B	Cyclic depsipeptide	*Haliclona sp.*	Inflammation

NRPs agents	Chemical class	Origin	Target
Perthamide C-D	Cyclopeptide	*Theonella swinhoei*	Inflammation
Solomonamide A- B	Cyclic peptide	*Theonella swinhoei*	Inflammation
Stylissatin A	Cyclic peptide	*Stylissa massa*	Murine macrophage RAW264.7
Dicynthaurin	—	*Halocynthia aurantium*	Antimicrobial
Nagahamide A	Depsipeptide	*Theonella swinhoei*	Antibacterial
Plicatamide	Octapeptide	*Styela plicata*	Antimicrobial
Callipeltins	—	*Latrunculia sp.*	*Antifungal/Candida albicans*
Citronamides A- B	—	*Citronia astra*	*Antifungal/Saccharomyces cerevisiae*
Renieramide	Cyclic tripeptide	*Reniera* sp.	—
Phoriospongins A-B	Depsipeptide	*Phoriospongia* sp. and *Callyspongia bilamellata*	Nematocidal/*Haemonchus contortus*

Table 4.
Agents produced from marine sponges and tunicates which are based on NRPs [7].

In the NCBI database, there are currently about 1.164 distinct non-ribosomal peptides that form over 500 different monomers including both proteinogenic and non-proteinogenic L- and D-amino acids, as well as amines and carboxylic acids. These complex secondary metabolites' linear, cyclic, branching, or other complicated primary structures are frequently altered to enhance clinical qualities and/or bypass resistance mechanisms. Indeed, the nucleotide sequence modification of a native NRPS gene or mixing modules from multiple NRPSs makes them more efficient with pharmacological properties. Several bioengineering and molecular techniques have been developed during the last few decades to produce modified NRPs with improved physicochemical characteristics and bioactivity [13].

5. Conclusion

In this chapter, we discussed the significance, synthesis, and application areas of NRPs-based agents, which have received a lot of interest as a new source of pharmaceutical agents. NRPs with unique chemical structures and diverse biological actions, such as antibacterials (penicillin, vancomycin), anticancer compounds (bleomycin), and immunosuppressants (cyclosporine), have been researched as novel compounds for new drug discovery and development throughout the last several decades. *In vitro* bioassays and the transfer of biosynthetic gene clusters of NRPs have been the focus of the majority of these studies. For the development of NRPs drugs with improved pharmacological properties, genetic manipulation and molecular approaches will allow the rapid construction of new NRPSs containing specific point mutations or exchanged domains.

Author details

Sadık Dincer[1]*, Hatice Aysun Mercimek Takci[2] and Melis Sumengen Ozdenefe[3]

1 Cukurova University, Adana, Turkey

2 Kilis 7 Aralık University, Kilis, Turkey

3 Near East University, Nicosia, Northern Cyprus

*Address all correspondence to: sdincer@cu.edu.tr

References

[1] Tsuge K, Matsui K, Itaya M. Production of the non-ribosomal peptide plipastatin in *Bacillus subtilis* regulated by three relevant gene blocks assembled in a single movable DNA segment. Journal of Biotechnology. 2007;**129**:592-603. DOI: 10.1016/j.jbiotec.2007.01.033

[2] Siewers V, San-Bento R, Nielsen J. Implementation of communication-mediating domains for non-ribosomal peptide production in *Saccharomyces cerevisiae*. Biotechnology and Bioengineering. 2010;**106**(5):841-844. DOI: 10.1002/bit.22739

[3] Zhang H, Liu Y, Wang X, Hu R, Xu G, Mao C, et al. Gene sequence diversity of the nonribosomal peptide and polyketide natural products in Changbaishan soil correlates with changes in landscape belts. Ecological Indicators. 2021;**133**:108160. DOI: 10.1016/j.ecolind.2021.108160

[4] Corpuz JC, Sanlley JO, Burkart MD. Protein-protein interface analysis of the non-ribosomal peptide synthetase peptidyl carrier protein and enzymatic domains. Synthetic and Systems Biotechnology. 2022;**7**(2):677-688. DOI: 10.1016/j.synbio.2022.02.006

[5] Martinez-Nunez MA, López y López VE. Nonribosomal peptides synthetases and their applications in industry. Sustainable Chemical Processes. 2016;**4**:13. DOI: 10.1186/s40508-016-0057-6

[6] Duban M, Cociancich S, Leclère V. Nonribosomal peptide synthesis definitely working out of the rules. Microorganisms. 2022;**10**:577. DOI: 10.3390/ microorganisms10030577

[7] Agrawal S, Adholeya A, Deshmukh SK. The pharmacological potential of non-ribosomal peptides from marine sponge and tunicates. Frontiers in Pharmacology. 2016;**7**:333. DOI: 10.3389/fphar.2016.00333

[8] Tippelt A, Nett M. *Saccharomyces cerevisiae* as host for the recombinant production of polyketides and nonribosomal peptides. Microbial Cell Factories. 2021;**20**(1):161. DOI: 10.1186/s12934-021-01650-y

[9] Izoré T, Candace Ho YT, Kaczmarski JA, Gavriilidou A, Chow KA, Steer DL, et al. Structures of a non-ribosomal peptide synthetase condensation domain suggest the basis of substrate selectivity. Nature Communication. 2021;**12**(1):2511. DOI: 10.1038/s41467-021-22623-0

[10] Fortinez CM, Bloudoff K, Harrigan C, Sharon I, Strauss M, Schmeing TM. Structures and function of a tailoring oxidase in complex with a nonribosomal peptide synthetase module. Nature Communications. 2022;**13**(1):548. DOI: 10.1038/s41467-022-28221-y

[11] Bozhuyuk KAJ, Fleischhacker F, Linck A, Wesche F, Tietze A, Niesert CT, et al. De novo design and engineering of non-ribosomal peptide synthetases. Nature Chemistry. 2018;**10**:275-281. DOI: 10.1038/NCHEM.2890

[12] Oestreich AM, Suli MI, Gerlach D, Fan R, Czermak P. Media development and process parameter optimization using statistical experimental designs for the production of nonribosomal peptides in *Escherichia coli*. Electronic Journal of Biotechnology. 2021;**52**:85-92. DOI: 10.1016/j.ejbt.2021.05.001

[13] Agrawal S, Acharya D, Adholeya A, Barrow CJ, Deshmukh SK. Nonribosomal

peptides from marine microbes and their antimicrobial and anticancer potential. Frontiers in Pharmacology. 2017;**8**:828. DOI: 10.3389/fphar.2017.00828

[14] Schwarzer D, Finking R, Marahiel MA. Nonribosomal peptides: From genes to products. Natural Product Reports. 2003;**20**(3):275-287. DOI: 10.1039/b111145k

[15] Kittilä T, Kittel C, Tailhades J, Butz D, Schoppet M, Büttner A, et al. Halogenation of glycopeptide antibiotics occurs at the amino acid level during non-ribosomal peptide synthesis. Chemical Science. 2017;**8**(9):5992-6004. DOI: 10.1039/c7sc00460e

[16] Wang H, Fewer DP, Holm L, Rouhiainen L, Sivonen K. Atlas of nonribosomal peptide and polyketide biosynthetic pathways reveals common occurrence of nonmodular enzymes. Proceedings of the National Academy of Sciences of the United States of America. 2014;**111**(25):9259-9264. DOI: 10.1073/pnas.1401734111

[17] Chang Z, Flatt P, Gerwick WH, Nguyen VA, Willis CL, Sherman DH. The barbamide biosynthetic gene cluster: A novel marine cyanobacterial system of mixed polyketide synthase (PKS)-non-ribosomal peptide synthetase (NRPS) origin involving an unusual trichloroleucyl starter unit. Gene. 2002;**296**(1-2):235-247. DOI: 10.1016/S0378 1119(02)00860-0

[18] Bozhueyuek KAJ, Watzel J, Abbood N, Bode HB. Synthetic zippers as an enabling tool for engineering of non-ribosomal peptide synthetases. Angewandte Chemie International Edition in English. 2021;**60**(32):17531-17538. DOI: 10.1002/anie.202102859

[19] Kries H, Niquille DL, Hilvert D. A Subdomain swap strategy for reengineering nonribosomal peptides. Chemistry & Biology. 2015;**22**:640-648. DOI: 10.1016/j.chembiol.2015.04.015

[20] Takahashi H, Kumagai T, Kitani K, Mori M, Matoba Y, Sugiyama M. Cloning and characterization of a *Streptomyces* Single module type non-ribosomal peptide synthetase catalyzing a blue pigment synthesis. The Journal of Biological Chemistry. 2007;**282**(12):9073-9081. DOI: 10.1074/jbc.M611319200

[21] Tooming-Klunderud A, Rohrlack T, Shalchian-Tabrizi K, Kristensen T, Jakobsen KS. Structural analysis of a non-ribosomal halogenated cyclic peptide and its putative operon from Microcystis: Implications for evolution of cyanopeptolins. Microbiology. 2007;**153**:1382-1393. DOI: 10.1099/mic.0.2006/001123-0

[22] Tapi A, Chollet-Imbert M, Scherens B, Jacques P. New approach for the detection of non-ribosomal peptide synthetase genes in Bacillus strains by polymerase chain reaction. Applied Microbiology and Biotechnology. 2010;**85**(5):1521-1531. DOI: 10.1007/s00253-009-2176-4

[23] Süssmuth RD, Mainz A. Nonribosomal Peptide Synthesis-Principles and Prospects. Angewandte Chemie International Edition in English. 2017;**56**:3770-3821. DOI: 10.1002/anie.201609079

[24] Zhou K, Zhang X, Zhang F, Li Z. Phylogenetically diverse cultivable fungal community and polyketide synthase (pks), non-ribosomal peptide synthase (nrps) genes associated with the South China Sea sponges. Microbial Ecology. 2011;**62**:644-654. DOI: 10.1007/s00248-011-9859-y

[25] Sieber SA, Marahiel MA. Molecular mechanisms underlying nonribosomal

peptide synthesis: Approaches to new antibiotics. Chemical Reviewes. 2005;**105**:715-738. DOI: 10.1021/cr0301191

[26] Dell M, Dunbar KL, Hertweck C. Ribosome-independent peptide biosynthesis: The challenge of a unifying nomenclature. Natural Product Reports. 2021:1-7. DOI: 10.1039/d1np00019e

[27] Iacovelli R, Bovenberg RA, Driessen AJ. Nonribosomal peptide synthetases and their biotechnological potential in *Penicillium rubens*. Journal of Industrial Microbiology and Biotechnology. 2021;**48**(kuab045). DOI: 10.1093/jimb/kuab045

[28] Vinothkumar S, Parameswaran P. Recent advances in marine drug research. Biotechnology Advances. 2013;**31**:1826-1845. DOI: 10.1016/j.biotechadv.2013.02.006